THE ART OF THE INTELLIGIBLE

THE WESTERN ONTARIO SERIES
IN PHILOSOPHY OF SCIENCE

A SERIES OF BOOKS
IN PHILOSOPHY OF SCIENCE, METHODOLOGY, EPISTEMOLOGY,
LOGIC, HISTORY OF SCIENCE, AND RELATED FIELDS

VOLUME 63

THE ART OF
THE INTELLIGIBLE

An Elementary Survey of Mathematics in its Conceptual Development

by

JOHN L. BELL
University of Western Ontario,
London, Canada

KLUWER ACADEMIC PUBLISHERS
DORDRECHT / BOSTON / LONDON

A C.I.P. Catalogue record for this book is available from the Library of Congress.

ISBN 0-7923-5972-0 (HB)
ISBN 1-4020-0007-3 (PB)

Published by Kluwer Academic Publishers,
P.O. Box 17, 3300 AA Dordrecht, The Netherlands.

Sold and distributed in North, Central and South America
by Kluwer Academic Publishers,
101 Philip Drive, Norwell, MA 02061, U.S.A.

In all other countries, sold and distributed
by Kluwer Academic Publishers,
P.O. Box 322, 3300 AH Dordrecht, The Netherlands.

Printed on acid-free paper

Printed in the Netherlands

To my dear wife Mimi

The purpose of geometry is to draw us away from the sensible and the perishable to the intelligible and eternal.

Plutarch

TABLE OF CONTENTS

FOREWORD

My purpose in writing this book is to present an overview—at a fairly elementary level—of the conceptual evolution of mathematics. As will be seen from the Table of Contents, I have adhered in the main to the traditional tripartite division of the subject into Algebra (including the Theory of Numbers), Geometry, and Analysis. I have attempted to describe, in roughly chronological order, what are to me some of the most beautiful, and—if I am not entirely misguided—some of the most significant developments in each of these domains. In this spirit I have also included brief accounts of mathematical notation, ancient Greek mathematics, set theory, and the philosophy of mathematics. The Appendices contain short expositions of topics which are particularly dear to my heart and which I hope my readers (if any) will take to theirs. My approach here—a curious mix, admittedly, of the chronological and the expository—is the result, not entirely of my whim, but also of having spent a number of years of lecturing on these topics to undergraduates.

I should point out that the use of the word "intelligible" in the book's title is intended to convey a double meaning. First, of course, the usual one of "comprehensible" or "capable of being understood." But the word also has an older meaning, namely, "capable of being apprehended only by the intellect, not by the senses"; in this guise it serves as an antonym to "sensible". It is precisely with this signification that Plutarch uses the word in the epigraph I have chosen. While the potential intelligibility of mathematics in this older sense is hardly to be doubted, I can only hope that my book conveys something of that intelligibility in its more recent connotation.

ACKNOWLEDGEMENTS

It was Ken Binmore—my colleague for many years at the London School of Economics—who proposed in the 1970s that I develop a course of lectures under the catch-all title "Great Ideas of Mathematics". Twenty years or so later Rob Clifton—my erstwhile colleague at the University of Western Ontario—suggested that I give the same course here, a stimulus which led to the expansion and polishing of the somewhat primitive notes I had prepared for the original course, and which also had the effect of emboldening me to turn them into the present book. In the last analysis it must be left to readers of the book to judge whether my two confrères are to be applauded for their encouragement of my efforts, but for my part I am happy to acknowledge my debt to them. I would also like to thank Elaine Landry for suggesting that I give the course of lectures on the philosophy of mathematics which ultimately led to the writing of Chapter 12 of the book. I am grateful to Alberto Peruzzi for his helpful comments on an early draft of the manuscript, and to Max Dickmann for his scrutiny of the hardback edition of the book, leading to the discovery of a number of errata (which are listed on a separate sheet in the present edition). To my wife Mimi I tender special thanks both for inscribing the Chinese numerals in Chapter 1 and for her heroic efforts in reading aloud to me the entire typescript (apart, of course, from the formulas). I must also acknowledge the authors of the many books—listed in the Bibliography—which I have used as my sources. Finally, I would like to record my gratitude to Rudolf Rijgersberg at Kluwer for the enthusiasm and efficiency he has brought to the project of publishing this book.

CORRIGENDA

p. 25, figure. "C" should be inserted in triangle NCS.

–, footnote. Add "r" at end of line.

p. 35, line –10. Add "$= 90 + r$" at end of formula.

p. 40, line 14. Insert space between "–" and "1".

p. 56, line 6 from bottom. The r.h.s. of the equation should read "$/[a + b + 2/(ab)]$".

p. 57, top line. Delete entire line.

–, line 9. For "r" read "r_1".

p. 62, line –6. Delete the second "to".

p. 64. Insert "=" after the second cube root.

p. 71, line –10. The third equation should read "$x(1 - x) = 0$".

p. 73. Delete two lines of footnote at bottom of page.

p. 74, line –3. For "questioned" read "question".

p. 75. Insert at top of page: "Quaternions may be regarded as describing *rotations* and *stretchings* (or *contractions*) in three-dimensional space. Given two vectors $v = a\mathbf{i} + b\mathbf{j} + c\mathbf{k}$ and $vN = aN\mathbf{i} + bN\mathbf{j} + cN\mathbf{k}$ we seek a quaternion $\mathbf{q} = w + x\mathbf{i} + y\mathbf{j} + z\mathbf{k}$ which when "applied" to v rotates or stretches it to coincide with vN in the sense that $vN = \mathbf{q}v$, that is,".

p. 76, line 13. Delete the commas.

p. 80, line 2. For "x_n' " read For "x_m' "

p. 91, line 18. Separate "generally" and "in".

p. 93, lines 12, 13. For "xy, where x and y" read "$x_1y_1 + \ldots + x_ny_m$, where $x_1, \ldots, x_n, y_1, \ldots, y_n$".

p. 111, line 11. For "*Methode*" read "*Méthode*"

p. 117, line 8. For "points differ", read "points of inflection differ".

p. 120, line 6. For "*Ilements*" read "*Iléments*".

p. 134, line 15. For "*Géometrie*" read "*Géométrie*".

–, line –3. For "any theorem", read "the dual of any theorem".

p. 137, bottom line. For "normal to the curve" read "normal to the surface".

p. 140, line 2. Delete second period at end of line.

p. 150, lines 6, 7. For "on the page following" read "below".

p. 152, line 4. At end of line add "$d(x, y) = 0$ if and only if $x = y$."

p. 157, bottom line. The equation should read "$S = 2\pi rh + 2\pi r5$".

p. 158, top line. For "$2rh + 2r5$" read "$2\pi rh + 2\pi r5$".

p. 205, line 15. Separate "twentieth" and "century".

p. 207, line –3. For "$a = 2$", read "$a = /2$".

p. 209, line 4. Insert period after "1987)".

p. 217, line 19. Interchange the expressions "$3 + 2$, i.e., 5" and "6".

p. 218, line 7. For "$\underline{m}_n$" read "m_n".

p. 219, line –7. Insert additional parenthesis after "$\neg T(s(x_1, x_1)$".

p. 220, line 5. Insert additional parenthesis after "$\neg T(s(m, m)$".

p. 226, line 9. Should read : "$(fg)(x + \varepsilon) = (fg)(x) + \varepsilon(fg)'(x) = f(x)g(x) + \varepsilon(fg)'(x),$"

–, line 10. Should read "$(fg)(x + \varepsilon) = f(x + \varepsilon)g(x + \varepsilon) = [f(x) + \varepsilon f'(x)].[g(x) + \varepsilon g'(x)]$".

p. 228, line 6. Should read "$k\varepsilon [T(x, t) + \frac{1}{2}\varepsilon (MT/Mx)(x, t)] = k\varepsilon T(x, t)$.".

CHAPTER 1

NUMERALS AND NOTATION

FOR CENTURIES MATHEMATICS WAS DEFINED as "the science of number and magnitude", and while this definition cannot nowadays be taken as adequate, it does, nevertheless, reflect the origins of mathematics with reasonable fidelity. Notions related to the concepts of number and magnitude can be traced back to the dawn of the human race. Indeed some animal species—whose origins antedate those of humanity by millions of years—behave in such a way as to reveal a rudimentary mathematical sense: experiments with crows, for example, have shown that birds can distinguish among sets containing up to four elements. At any rate, mathematical thinking has long played a role in the workaday life of human beings.

The origin of the *number concept*, in particular, would seem to lie in our remote ancestors' grasping the idea of *plurality*, and seeing that pluralities or collections of things can be both *matched* and *compared in size*. For example, the hands can be matched with the feet or the eyes, the fingers with the toes, but not the feet with the fingers. The realization that matchability of hands, feet, eyes, and any other pair of objects is independent of their nature must have provided a crucial stimulus for the emergence of the idea of number. That small numbers such as two and four played an early and important role in human thought is shown by their special position in the grammar of certain languages, for example Greek, in which a distinction is made between one, two, and more than two, and Russian, which uses one noun case with numbers up to four, and a different one for larger numbers.

Numbers are assigned to collections by means of the process of *counting*, that is, the procedure of matching the elements of a collection successively with the ascending sequence of numbers, or number names. The recognition that the procedure of counting "one, two, three, four, ..." can be performed *intransitively*, in other words, that when counting it is not necessary to be actually counting *something*, is likely to have been instrumental in establishing the universality of the number concept. Indeed, it has been suggested that the art of counting arose in connection with primitive religious ritual and that the counting or *ordinal* aspect of number preceded the emergence of the quantitative or *cardinal* aspect. Whatever its origin, the procedure of counting naturally imposes an *order* on numbers, and it must have been grasped very early on that this order corresponds faithfully to the relative sizes of the collections that numbers are used to count.

As the awareness of number developed, it became necessary to express or represent the idea by means of *signs*. The earliest and most immediate mode of representing numbers is by means of the fingers, which conveniently enable numbers up to ten to be represented. For larger numbers heaps of pebbles were used, piled in groups of five, each corresponding to the number of fingers on the hand. In this connection it is of interest to note that the term "calculate" derives from the Latin word *calculus* meaning "small stone". Counting by fives and tens—and, in some cultures, twenties—became standard practice, largely displacing the earlier systems of counting by twos and threes. Thus arose the idea of a *base* or *scale* for counting, in which some number *b* is selected, names for (some of the) numbers 1,..., *b* are assigned, and names for numbers larger than *b* are formulated as combinations of these. It was observed by Aristotle that the customary choice of base 10 is merely the result of the accidental fact that human beings happen to possess five fingers on each of two hands: from a strictly mathematical point of view it is somewhat regrettable that our ancestors did not possess a composite number of fingers, such as four or six, on each hand, rather than the awkward prime number five.

Heaps of pebbles, lacking ready portability as well as the requisite stability for prolonged storage of numerical information, came to be replaced in prehistory by less ephemeral and more portable devices, such as notches in a piece of wood or bone. The oldest known example of such a device is a tally stick dating from paleolithic times found in Moravia in 1937. It is the radius bone of a young wolf on which are incised 55 notches, arranged in two series, with 25 in the first and 30 in the second, both assembled in groups of five. The existence of these and a few other, similar, artifacts, being more than thirty thousand years old, show that the use of the number concept, and, in particular, that of the number five, long antedates such technological advances as metal smelting or wheeled vehicles, and even the development of written language. This latter fact suggests that the appearance of number *signs* preceded that of number *words*, a plausible claim in any case since primitive signs for numbers such as notches or strokes possess a visual immediacy which has no counterpart in speech. The primacy of the *sign* or *symbol* in mathematics has persisted until the present day.

The earliest *numerals* or formal signs for numbers of which we possess definite record appeared in Egypt around 3400 B.C. The Egyptian system of *hieroglyphic* numerals employed strokes for numbers below ten, and special symbols for powers of ten, for example:

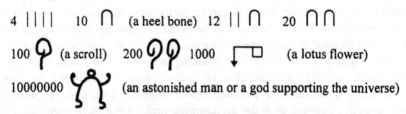

Numerals appeared in Egypt long before the first known numerical inscriptions in India (3rd century B.C.), China (3rd century B.C.) and Crete (1200 B.C.).

Egyptian numerals constitute what is known as a *simple grouping system*. In such systems some number *b*—called a *base* or *scale* (10 in the Egyptian, as in most other cases)—is chosen and the powers *b*, b^2, b^3, ... of *b* treated as units—the *higher units.* Symbols for the latter, as well as for 1, are introduced and arbitrary numbers then represented by deploying these symbols *additively*, each symbol being repeated the requisite number of times. Thus, for example, in Egyptian hieroglyphics written left to right[1],

$$3215 = \quad \text{⌐⌐ ⌐⌐ ⌐⌐ 𝒫 𝒫 ∩ |||||}$$

Another example of a simple grouping system is provided by the Babylonian (c.2000 B.C.) symbols for numbers below 60. The early Babylonians (Sumerians) used clay as a medium for writing, producing inscriptions—the so-called *cuneiform* characters—by means of a stylus with a wedge-shaped tip. On the resulting clay tablets numbers less than 60 were expressed by a simple grouping system to base 10. Here we find a novelty, for the inscriptions are often simplified by the use of a *subtractive* symbol, viz.,

The numbers 1 and 10 were symbolized by ⊳ and ◁ . Thus, for example,

$$24 = \quad \text{◁ ◁ ⧽ ⧽} \qquad 28 = 30 - 2 = \quad \text{◁ ◁ ◁ ∇ ⊳ ⊳}$$

As a further example of a simple grouping system we may consider the *Attic Greek* numerals developed some time prior to the 3rd century B.C. Here the powers of 10 are symbolized by the initial letters of number names; in addition there is a symbol for 5. These initial numerals, in modern notation, were

    Π (Γ), *pi*, for ΠΕΝΤΕ (pente) five
    Δ , *delta*, for ΔΕΚΑ (deka), ten
    H, an old breathing symbol, for HEKATON, hundred
    X, *chi*, for ΧΙΛΙΟΙ (chilioi), thousand
    M, *mu,* for ΜΥΡΙΟΙ (murioi), ten thousand

Thus, for example, they wrote

    ⌐Δ   pente-deka, for $5 \times 10 = 50$

    ⌐H   pente-hekaton, for $5 \times 100 = 500,$

---

[1] As in Semitic languages, the Egyptians generally wrote from right to left.

and so

$$3756 = XXX\lceil HHH\lceil \Delta\Gamma\rceil .$$

By far the most familiar simple grouping system is of course that based on the *Roman numerals*. Although nowadays it has only an ornamental use, the Roman system had the merit that the majority of its users needed to commit to memory the values of just four letters: V (5), X (10), L (50) and C (100). For larger numbers the symbols D (500) were employed. The simple grouping of symbols in the Roman system was eventually combined with the *subtractive principle*, in which a symbol for a smaller unit placed before a symbol for a larger one indicates that it is to be subtracted from the latter, as, e.g.

$$1949 = MCMXLIX.$$

The subtractive principle was employed only sparingly in ancient and medieval times, when the above number would have been written less succinctly

$$1949 = MDCCCCXXXXVIIII.$$

The question of the origin of the form of the Roman numerals has aroused considerable speculation. Of the various theories it is that of the German historian *Theodor Mommsen* (1817-1903) which has the strongest epigraphic support, and has gained the widest acceptance by scholars. Mommsen thought that the use of V for 5 resulted from its similarity in shape to the outspread hand with its 5 fingers, and that two of these gave the X for 10. He also contended that three of the other numerals were modified forms of Greek letters not employed in the Etruscan and early Latin alphabet. These were *chi*, which appears in inscriptions not only as **X**, but also in such forms as ⊥, ⌐, which were to become the symbol L which was arbitrarily chosen to denote 50; *theta*, Θ, chosen to represent 100, and which, through the influence of the word *centum* (100), finally became transformed into C; and *phi*, Φ, to which was assigned the value 1000, and whose written form (|) gradually became, under the influence of the word *mille* (1000), the symbol M. With delightful literalness, half of (|) , i.e., |) , was taken to represent 500 and was later written as D.

Roman numerals were poorly adapted for use in calculation, and actual computations were carried out on an abacus (from Greek *abax*, "slab"). Originally taking the form of sand or dust covered boards, by the late Roman period these devices had evolved into bronze tablets containing a number of grooves in which fixed counters slide. In each column a lower counter represents one, and the upper counter five, higher units.

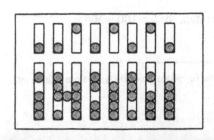

Multiplicative grouping systems are refinements of simple grouping systems. In a multiplicative grouping system, the selection of a base b and symbols for 1 and the higher units b, b^2, b^3,... is augmented by a set of symbols for 2, 3, ..., $b-1$: these latter, together with 1 are known as *digits*. The symbols of the two sets are then employed *multiplicatively* to indicate the number of higher units required to form a given number. Thus, if we designate the first nine numbers by the usual symbols 1, ..., 9 but agree to designate 10,100,1000 by u, v, w, say, then in the corresponding multiplicative system we would write, for example

$$7843 = 7w8v4u3.$$

The traditional Chinese numeral system is an example of such a system to base 10. Writing vertically, the symbols of the two basic groups and the representation of the number 7843 are

Although concrete historical evidence is lacking, it may be surmised that at some point users of multiplicative grouping systems came to see that the values of the higher units could be indicated *locally*, that is, by the mere *position* of the digits, thereby enabling the symbols for higher units to be dispensed with altogether. But in so doing ambiguities will arise unless a new digit—a *zero*—is introduced to signalize powers of the base that happen to be missing in the representation of a given number. (For instance, without such a symbol we would be unable to distinguish, say, 22 from 202.) Thus, in a *positional* number system on a base b, the symbols for the higher units are discarded and the digits augmented by a symbol for zero; each number N can then be uniquely expressed as a sum

$$N = a_n b^n + a_{n-1} b^{n-1} + ... + a_2 b^2 + a_1 b + a_0,$$

where a_0, ..., a_n are digits, so enabling N to be presented in the succinct form of a sequence of digits

$$N = a_n a_{n-1}...a_2 a_1 a_0.$$

The earliest number system incorporating the positional principle was developed by the Babylonians before 2000 B.C.: here numbers exceeding 60 were written *sexagesimally*, that is, positionally in base 60. Lacking a symbol for zero, however, the system inevitably suffered from ambiguity. By 200 B.C. the Babylonians' successors had introduced a symbol—a proto-zero, so to speak—to indicate the absence of a number, but did not employ it in calculation, showing that they had not yet taken the crucial step of ascribing it full digithood.

The first fully positional system of enumeration was developed by the Mayas of Yucutan around 300 A.D. Scholars have come to believe that the large numbers inscribed on their monuments and in their few surviving bark-paper codices are the dates of lunar and solar eclipses presented in terms of the "Long Count", a cyclic calendar of religious origin apparently dating back many centuries. The Maya system is remarkable in its employment of a *mixed base* in addition to a symbol for zero: although essentially *vigesimal*, that is, based on 20, the third digit position is 18.20 = 360 instead of $20^2 = 400$ and subsequent digit positions are of the form 18.20^n. This curiosity almost certainly reflects the calendric origins of their numeration system, for the official Maya year had 360 days. The Mayas employed just three symbols: a dot for the unit, a bar for five, and a stylized shell for zero. Numbers below twenty were notated according to a simple grouping scheme

1	•	6	$\underline{\bullet}$	11	$\underline{\overset{\bullet}{}}$	16	$\underline{\overset{\bullet}{}}$
2	••	7	$\underline{\bullet\bullet}$	12	••	17	••
3	•••	8	•••	13	•••	18	•••
4	••••	9	••••	14	••••	19	••••
5	⎯	10	⎯⎯	15	⎯⎯⎯	0	⬭

Larger numbers were written vertically, for example:

$$43483 = 6.18.20^2 + 0.18.20 + 14.20 + 3 \ =$$

The system of numeration used almost universally today is the *Hindu-Arabic decimal system*, believed by scholars to have been introduced in India sometime before the ninth century A.D. and subsequently developed by Arab scholars who later transmitted it to the West. Not the least of its advantages is the fact that it enables calculations to be performed simply, in a purely symbolic manner, rendering unnecessary the use of an auxiliary device such as an abacus. In fact, the obvious correspondence between the lines of an abacus and the digit positions in the Hindu-Arabic representation of numbers suggests that the system may have originated as a symbolic representation of the abacus itself: this would make natural the introduction of a symbol for zero as corresponding to a line on the abacus empty of counters. In any event, by the sixteenth century the "algorists"—champions of the Hindu-Arabic system—had achieved supremacy in Europe over their rivals the "abacists"—those who still employed Roman numerals for the recording of numbers and the abacus for computation[2].

Once the Hindu-Arabic system had been fully accepted other advantages came to light, for instance, the fact that it provides a convenient notation for *fractions*, using inverse powers of 10. For example, 3/8 is represented by the *decimal*

$$0.375 = 3 \times 10^{-1} + 7 \times 10^{-2} + 5 \times 10^{-3}.$$

Of course, in this system some fractions will be represented by nonterminating (periodic) decimals, e.g.

$$\tfrac{1}{3} = 0.33333.... = 3 \times 10^{-1} + 3 \times 10^{-2} + 3 \times 10^{-3} + ...,$$

in which the sum on the right approaches ⅓ arbitrarily closely as more terms are "added in". Expressing fractions in decimal notation enables calculations with them to be performed in exactly the same way as with whole numbers.

The actual form of the Hindu-Arabic numerals underwent considerable variation before being finally stabilized by the invention of printing in the fifteenth century. It has been suggested that the numerals are of *iconic* origin: specifically, that they were originally formed by drawing as many strokes or dots as there are units

---

[2] It is an irony that the development of the electronic computer —the abacus of today—is rapidly reducing the role of the Hindu-Arabic numeral system to that of its Roman counterpart, i.e., to a device for the mere recording of numbers. But this "abacists' revenge" has come at a price, since, unlike the lines of an abacus, the internal structure of a computer is far too complex to suggest the form of *its* immediate successor.

represented by the respective numerals, these strokes or dots later becoming connected through the use of cursive writing. While this theory does not seem entirely implausible (at least in the case of 1, 2, and 3), it is not supported by solid historical evidence, nor in any case does it explain the great variety of forms which the numerals took at different times and places.

The symbol "0" for zero may have evolved from the dot "•" which was first used by the Hindus. Before definitively assuming its present form, the symbol "0" was sometimes crossed by a horizontal or vertical line, causing it to resemble the Greek letters "Θ" or "Φ". The word "zero" itself seems to have come through Old Spanish from the Arabic *sifr*, "empty", from which our word *cipher* also derives.

Greater confidence may be placed in theories accounting for the form of the symbols for *operations* on numbers, such as addition and multiplication, since these are of considerably more recent origin than the signs for the numbers themselves. The modern algebraic symbols "+" and "−" for addition and subtraction first appear in fifteenth-century German manuscripts. It is highly likely that the "+" sign descended from a cursive form of *et*, "and", in Latin manuscripts. Concerning the origin of the minus symbol "−" nothing certain is known, but it may simply have arisen as a sign indicating separation. The sign "×" (St. Andrew's cross) for multiplication first appears in *William Oughtred's* (1574–1660) *Clavis Mathematicae* of 1631, but it seems likely that he derived it from earlier uses of the cross in solving problems involving proportion. Thus, for example,

was used to indicate that 3 was to be multiplied by 3, and 4 by 2. The use of the dot "·" for multiplication[3] was explicitly introduced by the German mathematician and philosopher *Gottfried Wilhelm Leibniz* (1646–1716) in 1698 in a letter to John Bernoulli, in which he states: "I do not like × as a symbol for multiplication as it is easily confounded with *x*... often I simply relate two quantities by an interposed dot and indicate multiplication by ZC · LM." The sign "÷" for division was originally employed for subtraction; its first recorded use as a sign for division (deriving perhaps from the fractional line "—") is in a Swiss Algebra of 1659. Its use then quickly spread to the English speaking countries, but not to the rest of continental Europe, where Leibniz's symbol ":" for division remained in force well into the twentieth century.

The familiar sign "=" for equality was introduced by the English mathematician *Robert Recorde* (c. 1510–1558) in his charmingly titled algebra treatise

---

[3] It is interesting that in algebra multiplication is indicated by *juxtaposition*: thus *ab* stands for "*a* times *b*." This may derive from the fact that in everyday language number terms can be used adjectivally, as in "four apples"; when the modified noun is itself a number, the resulting phrase denotes a multiplication: for example, "four threes" is taken to mean "four times three", i.e. 12. And indeed the multiplication tables were (until recently) learned precisely by repeating phrases of this sort. Juxtaposition is of course systematically employed in positional number systems to indicate *addition* of numerals, and this probably explains why the same device is employed to indicate the adding of a fraction to an integer: thus, for example, "5¾" stands for "5 + ¾".

The Whetstone of Witte, published in 1557. Recorde says he was led to adopt a pair of equal parallel line segments as the symbol for equality "bicause noe 2 thynges can be more equalle." Despite its simplicity and suggestiveness, Recorde's sign did not gain immediate acceptance; in fact for two centuries a struggle for supremacy took place between it and *René Descartes'* (1596–1650) sign " ∞ " — possibly derived from the first two letters of the Latin *aequalis*, "equal"— which he had introduced in his *Géométrie* of 1637. The final victory of "=" at the close of the seventeenth century seems to have been mainly due to the influence of Leibniz, who, with his usual eye for the appropriate, had adopted Recorde's symbol.

The signs ">" and "<" for "greater" and "less" were introduced by the English mathematician *Thomas Harriot* (1560–1621) in his work *Artis Analyticae Praxis* of 1631. Like Recorde's sign for equality, despite their simplicity and suggestiveness, the use of these signs spread only rather slowly, not becoming standard until the nineteenth century.

CHAPTER 2

THE MATHEMATICS OF ANCIENT GREECE

THE CULTURE OF ANCIENT GREECE (600 B.C. – 300 A.D.) still astonishes today by its originality, depth and diversity. The works of their philosophers Plato and Aristotle, of their tragedians Euripides and Sophocles, of their poets Homer and Pindar, of their historians Herodotus and Thucydides—and many others—have had an incalculable influence on the development of the arts. No less influential were the scientists and mathematicians of ancient Greece: for centuries the name "Euclid" has been inseparable from the term "geometry" and mathematics known as "the Greek science".

Mathematics—or to be more precise, *geometry*, a word which derives from the Greek *geometrein*, "earth measurement"—begins in Greece with *Thales of Miletus* (*c*.624 – 547 B.C.), whose travels in Egypt are believed to have furnished the original sources of his geometric knowledge. Egyptian geometry itself seems to have arisen from the practical requirement of fixing accurately the boundaries of landed property subject to periodic flooding by the river Nile. As attested by the Rhind papyrus (*c*.1700 B.C.) and other documents, Egyptian geometry consisted in the main of practical rules for measuring areas of figures such as squares, triangles, circles, etc. and volumes of measures of corn, grain, etc. of various shapes. It may be inferred from the fact that the Egyptians constructed pyramids of definite slope that they were also in possession of the idea of similarity of figures, and certainly of triangles. Moreover, the Egyptians knew that a triangle with its sides in the ratio 3:4:5 is right-angled, and made use of this fact as a means of drawing right angles. However, nothing indicates that they were aware of the general property linking the sides of a right-angled triangle, or that they proved any general geometric theorem.

It may be plausibly surmised that the diagrams illustrating the measurement of various plane figures that Thales would have seen during his Egyptian travels suggested to him the idea of isolating the general principles behind their construction. This in turn probably led to the discovery of the following theorems with which tradition associates him:

1. A circle is bisected by any diameter.
2. The base angles of an isosceles triangle are equal.
3. If two straight lines cut one another, the vertically opposite angles are equal.
4. The angle in a semicircle is a right angle.

Elementary as these assertions are, their *generality* already places them far beyond their Egyptian origins, and indeed marks them as the first steps towards a systematic *theory* of geometry.

After Thales came *Pythagoras of Samos* (572–497 B.C.) and his school, who are said to have coined the term "mathematics" from a root meaning "learning" or "knowledge". The remarkable advances in mathematics made by the Pythagoreans led them to the belief that mathematics, and more especially *number*, lies at the heart of all existence -- the first "mathematical" philosophy. For the Pythagoreans the structure of mathematics took the form of a bifurcating scheme of oppositions:

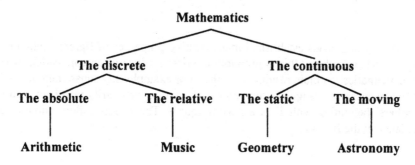

This scheme is the source of the so-called *quadrivium*, which served as the basis for Western pedagogy until the end of the Middle Ages. The further addition of the *trivium*, comprising grammar, rhetoric and logic, formed what came to be known as the *seven liberal arts*.

The Pythagoreans elevated the idea of number above the realm of practical utility and transformed it into a pure concept, creating what we may call the *Pythagorean theory of numbers*. This theory included definitions of the various classes of numbers: odd, even, prime, composite, odd times even, even times even, etc. Also ascribed to the Pythagoreans is the concept of *perfect* number, i.e. a number such as 6 or 28 which is the sum of its proper divisors. They were certainly the originators of the idea of *figurate* number, a link between number and geometry which must have played an important role in the creation of their numerical philosophy. Figurate numbers arise as the number of dots in certain geometric figurations such as triangles, squares, etc. For example, the triangular numbers appear as:

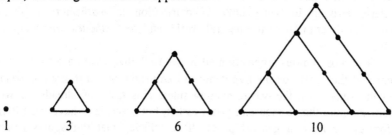

and the square numbers as:

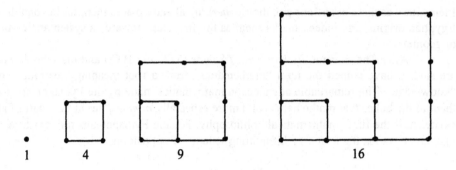

1   4   9   16

The Pythagoreans established many striking properties of figurate numbers in a purely geometric fashion: this is probably the first instance of an application of one area of mathematics to obtain results in another. For example, any square number is the sum of two successive triangular numbers and the sum of arbitrarily many consecutive odd numbers, beginning with 1, is a perfect square. These facts follow immediately from a glance at the figures:

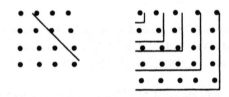

The fourth triangular number 10, the so-called "tetraktys", played an important role in the Pythagorean scheme as the sum of the possible geometric dimensions: 1 point, the generator; 2 points, a line; 3 points, a plane; 4 points, a (tetrahedric) volume.

The Pythagoreans are also credited with discovering the correspondence between musical intervals and simple arithmetic ratios of lengths of stretched strings: the octave corresponding to the ratio 2:1, the fifth to 3:2 and the fourth to 4:3. This discovery, which later led to the construction of musical modes and scales, is particularly striking in being both a contribution to *mathematical physics*, the mathematical description of the natural world, and to *aesthetics*, the analysis of the beautiful.

The most famous proposition attributed to Pythagoras himself is the geometric theorem that the square on the hypotenuse of any right-angled triangle is equal to the sum of the squares on the two remaining sides. This fact was already known to the Babylonians more than a thousand years earlier, but it seems to have been Pythagoras who formulated its first general proof. It is believed that Pythagoras's proof was *dissective* in nature, as suggested by the equal square figures below. The left hand figure is the sum of four copies of the original right triangle with the sum of the squares on the legs, and the right hand figure is the sum of four such copies with the square on the hypotenuse.

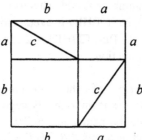 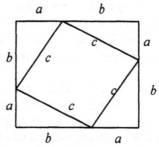

As we have said, it was a fundamental principle of Pythagoreanism that all is explicable in terms of properties of, and relations between, whole numbers—that number, indeed, forms the very essence of the real. It must, accordingly, have come as a great shock to the Pythagoreans to find, as they did, that this principle *cannot be upheld within geometry itself.* This followed upon the shattering discovery, probably made by the later Pythagoreans before 410 B.C., that ratios of whole numbers do not suffice to enable the diagonal of a square or a pentagon to be compared in length with its side. They showed that these line segments are *incommensurable,* that is, it is not possible to choose a unit of length sufficiently small so as to enable both the diagonal of a square (or a pentagon) and its side to be measured by an integral number of the chosen units. Pythagorean geometry, depending as it did on the assumption that all line segments are *commensurable* (i.e., measurable by sufficiently small units), and with it the whole Pythagorean philosophy, was dealt a devastating blow by this discovery. It may be remarked that the Pythagorean catastrophe shows how risky it can be to base a world outlook on a literal interpretation of mathematics: the exactness of mathematics may be used to undermine it!

The initial steps of what may have been the Pythagorean proof of the incommensurability of the side and diagonal of a square are presented in Plato's dialogue the *Meno,* in which Socrates asks an uneducated slave how to double the area of a square while maintaining its shape. Eventually Socrates solves the problem by producing the figure below, in which the inscribed square BDIH has double the area of the given square ABCD:

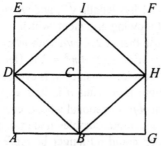

Having achieved his purpose, Socrates stops here, but it is likely that the Pythagoreans continued the argument along the following lines. Suppose AB and BD are commensurable; then we may take a unit of measure which divides both AB and DB an integral number of times, and, by successively doubling the length of the unit, if necessary, we may assume that at least one of these numbers is odd. Then (the area of) □AGFE is divisible by 4. But □AGFE = 2 □BHID, so □BHID is even, and so BD^2 and BD are even. Therefore □BHID is divisible by 4. But □BHID = 2 □ABCD so □ABCD is even. Hence AB^2, and AB are even. Contradiction.

It is possible that the actual *discovery* of the phenomenon of incommensurability was a by-product of the procedure—known to the Pythagoreans — of *subdividing a regular pentagon*. Starting with a regular pentagon *ABCDE*, and then drawing all five diagonals, we find that these intersect in points *A'B'C'D'E'* which form another regular pentagon, whose diagonals in turn form a still smaller regular pentagon, and so on. Clearly this process of subdivision never stops, and there is no "smallest" pentagon whose side could serve as a unit of measure. To be precise, let *AB* = *a*, *AC* = *d*, *A'B'* = *a'*, *A'C'* = *d'*. Then $d' = d - a < 1/2d$, $a' = 2a - d$. Therefore, if *AB* and *AC*

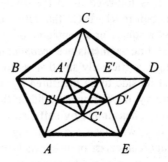

were commensurable by some unit of length *u*, the side and diagonal of all smaller pentagons obtained by subdivision as above would be commensurable by the same unit *u*. Since these pentagons can be made arbitrarily small, and, in particular, so small that their sides are shorter than *u*, we have a contradiction.

In modern terms, the Pythagorean incommensurabilities are understood as asserting that the "numbers" representing the lengths of certain lines are *irrational*, that is, are not rational fractions of the form *m/n* with integral *m,n*. Thus, for example, the diagonal of a square of side 1 has length $\sqrt{2}$, and we may establish the irrationality of the latter by means of the following argument, which is essentially an algebraic version of one given above. Suppose that $\sqrt{2} = m/n$ with *m/n* in lowest terms so that at least one of *m,n* is odd. Then $2 = m^2/n^2$ so that $m^2 = 2n^2$. Hence m^2 is even, and so therefore is *m*. It follows that *n* must be odd. But since *m* is even, m^2 is divisible by 4, so that $n^2 = m^2/2$ is even and so therefore is *n*. Thus *n* is both even and odd, a contradiction. A similar argument shows that $\sqrt{p}$ is irrational whenever *p* is not a perfect square.

The discovery of incommensurables undermined the Pythagorean concept of "ratio" of magnitudes, which could no longer be represented in all cases in terms of whole numbers. It was *Eudoxus of Cnidus* (*c*.400–350 B.C.) who, by formulating an *axiomatic* theory of ratios, or proportions, resolved the problem of incommensurables as it affected geometry and in so doing anticipated the modern theory of real numbers.

In Eudoxus's theory, we are given a collection of *similar magnitudes*, e.g. line segments, or planes , or volumes, or angles, etc., together with the notion of *ratio* of similar magnitudes satisfying certain *axioms*, of which the following, expressed in modern terms, are the most important:

P₁. Given any two ratios, there is an integral multiple of the one which exceeds the other.

P₂. The ratio $a{:}b$ of two magnitudes a, b is equal to the ratio $c{:}d$ of two other magnitudes c, d if and only if, for any natural numbers m,n, na is greater than, equal to, or less than mb according as nc is greater than, equal to, or less than md.

Axiom **P₁**, which is also known as *Archimedes' principle*, has the effect of excluding infinitely small or infinitely large quantities, and axiom **P₂** is, in essence, a *definition of equality* for ratios.

Let us use these principles to prove the familiar proposition that *the areas of triangles having the same altitude are to one another as their bases*. The Pythagorean proof of this theorem assumed that the bases of the triangles are commensurable, but the Eudoxan proof succeeds in avoiding this assumption, as we shall see. Thus suppose given two triangles ABC and ADE, whose bases BC and DE lie on the same line MN. On CB produced, mark off, successively from B, $m-1$ segments equal to CB, and connect the points of division B_2, B_3, ..., B_m with vertex A as in the figure below. Similarly, on DE produced, mark off, successively from E, $n-1$ segments equal to DE, and connect the points of division E_2, E_3, ..., E_n with vertex A. Then $B_mC = m.BC$, $\triangle AB_mC = m.\triangle ABC$, $DE_n = n.DE$, $\triangle ADE_n = n.\triangle ADE$. Also, since the triangles AB_mC and

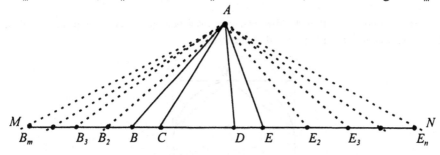

ADE_n have the same altitude, their areas will be greater, smaller, or coincide according as the lengths of their bases are greater, smaller, or coincide. It follows that $m.\triangle ABC$ is greater than, less than, or equal to $n.\triangle ADE$ according as $m.BC$ is greater than, less than, or equal to, $n.DE$. So, by **P₂**, $\triangle ABC : \triangle ADE = BC : DE$, and the proposition is established.

In addition to its major function of providing a rigorous foundation for the theory of similar figures in geometry, the Eudoxan theory of proportion played an important role in the justification of arguments based on the so-called *method of exhaustion*—in essence, an anticipation of the *integral calculus*—for computing the areas of regions bounded by curves, e.g. the circle. One of the earliest contributions to the problem of *squaring the circle*, that is, of constructing a square equal in area to a given circle, was made by *Antiphon the Sophist* (480–411B.C.) and *Bryson of Heraclea* (b. *c.*450 B.C.) , both contemporaries of Socrates. It is believed that their idea was to double successively the sides of a regular polygon inscribed in the circle, so that the difference in area between the polygon and the circle will eventually be exhausted. Since a square can be constructed equal in area to any given polygon, the squaring of the circle will be effected. This procedure met with the objection that, since magnitudes are divisible without limit, the area of the inscribed polygon will never in fact coincide with that of the circle. Nevertheless, it contains the germ of the method of exhaustion.

As an illustration of the method we give Eudoxus's proof that the area of a circle is directly proportional to the square on its diameter. Eudoxus bases his argument on the following assumptions:

M_1. Magnitudes are divisible without limit.
M_2. If from any magnitude there be subtracted a part not less than its half, from the remainder a part not less than its half, etc., then there will remain eventually a magnitude less than any given magnitude of the same kind.

Axiom M_1 expresses the idea that magnitudes are *continuous* entities and are not built from atoms. Axiom M_2 is known as *Eudoxus's principle of convergence*.

Now it must be shown that, if A_1 and A_2 are the areas of two circles with diameters d_1 and d_2, then

$$A_1: A_2 = d_1{}^2: d_2{}^2.$$

To do this we first show that the difference in area between a circle and an inscribed regular polygon can be made as small as desired.

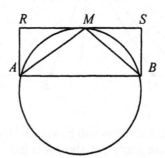

Let AB be a side of a regular inscribed polygon and let M be the midpoint of the arc AB. Since $\triangle AMB = \frac{1}{2} \square ARSB > \frac{1}{2} \cap AMB$, it follows that by doubling the number of sides of the regular inscribed polygon we increase the area of the polygon by more than half the difference in area between the polygon and the circle. Consequently, using M_1 and M_2 we see that by doubling the number of sides of the polygon sufficiently often, we can make the difference in area between it and the circle less than any assigned area, however small.

Now suppose, for contradiction's sake, that we had

$$A_1 : A_2 > d_1^2 : d_2^2.$$

Then, by the argument above, we can inscribe in the first circle a regular polygon whose area P_1 differs so little from A_1 that

$$P_1 : A_2 > d_1^2 : d_2^2.$$

Let P_2 be the area of a regular polygon similar to the original regular polygon, only inscribed in the second circle. Then, by elementary geometry,

$$P_1 : P_2 = d_1^2 : d_2^2,$$

so that

$$P_1 : A_2 > P_1 : P_2,$$

whence $P_2 > A_2$, an absurdity. A similar argument shows that we cannot have $A_1 : A_2 < d_1^2 : d_2^2$, and so we conclude that

$$A_1 : A_2 = d_1^2 : d_2^2,$$

as required.

It follows that, if A is the area and d the diameter of a circle, then $A = kd^2$, where k is a constant (by definition, equal to $\frac{1}{4}\pi$) for all circles.

The best known Greek mathematical text is the celebrated *Elements* of *Euclid*, written about 300 B.C. This definitive geometric treatise appears to have quickly and completely superseded all previous works of its kind: in fact, no trace whatsoever remains of its predecessors. With the exception of the Bible, no work has been circulated more widely or studied more assiduously, and no work, without exception, has exercised a greater influence on scientific thinking. Over a thousand editions of the *Elements* have appeared since the first was printed in 1482, and for more than two millenia it has dominated the teaching of geometry.

Euclid's *Elements* is not devoted to geometry alone, but also contains considerable number theory and elementary algebra. Although some of its proofs and propositions were doubtless invented by Euclid himself, the work is, in the main, a systematic compilation of the works of his predecessors. Euclid's achievement lies in his skilful selection of propositions and their arrangement in a logical sequence, creating in the process a standard of rigorous presentation that was not to be surpassed until modern times. Through its use of the *postulational-deductive*, or *axiomatic*, method the *Elements* became the universally accepted model for rigorous mathematical demonstration: it may be said without exaggeration that it is to the influence of this work that we owe the pervasiveness of the axiomatic method in the mathematics of today.

Applied to a given discourse, the essence of the axiomatic method is to begin by laying down a number of *definitions*, and then to introduce a group of statements — the *axioms* or *postulates* — which are accepted without proof. From these latter all the remaining true statements of the discourse are to be derived in a rigorous manner. Axioms and postulates were distinguished by Greek mathematicians and philosophers in at least three different ways:

(1) an axiom is a self-evident assertion about, and a postulate a construction of, a certain thing;

(2) an axiom is a universal assumption, whereas a postulate is an assumption germane only to the subject under study;

(3) an axiom is obvious and acceptable, while a postulate is not necessarily so.

It seems likely that in his *Elements* Euclid inclined toward the second of these distinctions.

As it has come down to us, the *Elements* opens with 23 definitions, of which the following may be taken as typical:

1. A *point* is that which has no part.
2. A *line* is breadthless length.
3. A *straight line* is a line which lies evenly with the points in itself.
8. A *plane angle* is the inclination to one another of two lines in a plane which meet one another and do not lie in a straight line.
10. When a straight line set up on a straight line makes adjacent angles equal to one another, each of the equal angles is *right*, and the straight line standing on the other is called *perpendicular* to that on which it stands.
15. A *circle* is a plane figure enclosed by one line such that all the straight lines falling upon it from one point among those lying within the figure are equal to one another.
23. *Parallel* straight lines are those which, being in the same plane and being produced indefinitely in both directions, do not meet one another in either direction.

It continues with five axioms and five geometric postulates:

A1. Things which are equal to the same thing are equal to each other.
A2. If equals be added to equals, the wholes are equal.
A3. If equals be subtracted from equals, the remainders are equal.
A4. Things which coincide with one another are equal to one another.
A5. The whole is greater than the part..

P1. It is possible to draw a straight line from any point to any other point.
P2. It is possible to produce a straight line indefinitely in that straight line.
P3. It is possible to describe a circle with any point as center and with a radius equal to any finite straight line drawn from the center.
P4. All right angles are equal to one another.

P5. If a straight line intersects two straight lines so as to make the interior angles on one side of it together less than two right angles, these straight lines will intersect, if indefinitely produced, on the side on which are the angles which are together less than two right angles.

From these ten statements Euclid proceeds to derive 465 geometric propositions, assembled into thirteen books. In the first few books are developed the basic properties of familiar plane figures such as triangles, squares and circles, and the construction, using straightedge and compasses, of regular polygons. In Book V we find a masterly exposition of Eudoxus's theory of proportions, which in Book VI is applied to plane geometry, yielding the fundamental theorems on similar triangles. Books VII, VIII and IX deal with elementary number theory. Book VII contains in particular an account of the process, known today as the *Euclidean algorithm*, for finding the greatest common divisor of a pair of numbers. (That is, divide the larger of the two numbers by the smaller. Then divide the divisor by the remainder. Continue the process of dividing the last divisor by the last remainder until the division is exact. The final divisor is the sought greatest common divisor.[1]) Book IX includes a number of significant propositions, in particular one equivalent to the so-called *fundamental theorem of arithmetic*, namely that any integer greater than 1 can be uniquely expressed as a product of *prime numbers*, that is, numbers not possessing any smaller divisors apart from 1). Another is a proof of the remarkable proposition that there are *infinitely many prime numbers*. (Given any number n, consider the number $(1.2.3....n) + 1$. This number is not divisible by n or any smaller number, so its prime factors must all exceed n. Thus, given any number, we can find a prime number which exceeds it, and so there must be infinitely many prime numbers.) Book X deals with incommensurable line segments and Eudoxus's method of exhaustion. The last three books are devoted to solid geometry and the determination of volumes. The final proposition is the beautiful result that there are exactly five *regular solids*, that is, polyhedra all of whose faces are congruent and all of whose vertex angles are equal, namely, the regular tetrahedron (with 4 faces), cube (6), octahedron (8), dodecahedron (12), and icosahedron (20).

The mode of presentation of the *Elements* is *synthetic* in that it passes from the known and simpler to the unknown and more complex, a procedure which has become standard in mathematical exposition. But the reverse process—*analysis*—must have played a significant role in the *discovery* of the theorems, even if scarcely any trace of it is to be found in the text itself.

While the *Elements* was regarded as the *ne plus ultra* of rigour well into the nineteenth century, a few mathematicians recognized that the work is marred by a number of logical deficiencies. One of these—of which Euclid himself seems to have been aware—is the use of superposition to establish the congruence of figures. Against this two objections may be raised. First, it involves the idea of *motion* for which no logical basis has been provided, and secondly, it presupposes without justification that a figure's properties remain unaltered when moved from one position to another. This is a very strong assumption concerning physical space. Another deficiency is the

---

[1] The Euclidean algorithm is discussed in more detail in Chapter 4.

vagueness of many of the Definitions, for example, those of point, line and surface. Other shortcomings stem from assumptions concerning the *continuity* of lines and circles which, although evident from the properties of figures, are not furnished with logical justification. Such is the case, for instance, in the proof of the very first proposition, which shows that an equilateral triangle can be erected on any line segment. It requires that two particular circles have a point in common, a claim which, while intuitively obvious on grounds of continuity (since the two curves "cross"), is not logically derivable from the initial assumptions. Euclid also makes frequent use of the assumption that straight lines crossing one another must have a common point.

It is implicit in Euclid's postulates that in constructing straight lines and circles the straightedge and compasses are to be employed in strict accordance with certain rules, namely: *with the straightedge one is permitted to draw a straight line of arbitrary length through any two distinct points; with the compasses one is permitted to draw a circle with any given point as centre and passing through any second given point.* These instruments, used in accordance with the rules just stated, have become known as *Euclidean tools*. Note that the straightedge is taken to be *unmarked*, and the compasses are not to be employed as dividers, i.e. for transferring distances[2].

The ancient Greek mathematicians knew that many geometric constructions, for example, bisecting angles, erecting perpendiculars, and extracting square roots, could be carried out using Euclidean tools. However, try as they might, they were unable to devise constructions, using Euclidean tools alone, for resolving the following problems—the *Three Famous Problems of Antiquity*—whose refractory nature has given them great mathematical notoriety.

1. *Doubling the cube*: the problem of constructing the edge of a cube having twice the volume of a given cube.

2. *Trisecting the angle*: the problem of dividing an arbitrary given angle into three equal parts.

3. *Squaring the circle:* the problem of constructing a square having an area equal to that of a given circle.

In modern terms, the first of these problems may be stated as that of constructing a line of length $\sqrt[3]{2}$, and the third as that of constructing a line of length π. Many attempts were made to solve them, but it was not until the nineteenth century, more than two thousand years after their formulation, that they were finally shown to be insoluble by Euclidean means[3]. The unremitting search by Greek mathematicians for solutions to these problems had a profound influence on Greek geometry and led to many important discoveries, of conic sections (see below) to name but one. In our era

---

[2] Remarkably, it is possible to show that dividers can be "reconstructed" using Euclidean tools alone.

[3] The insolubility of problem 3 is epitomized in the following word palindrome: *You can circle the square, can't you, but you can't square the circle, can you?*

the analysis of these problems has also had a fruitful influence, especially on the development of algebra.

Tradition has it that the problem of doubling the cube arose as the result of attempting to carry out the instructions of the oracle to Apollo at Delos. It is reported that an Athenian delegation was despatched to this oracle to enquire how a plague then raging could be curbed, to which the oracle responded that the cubic altar to Apollo must be doubled in size. The Athenians are said to have then dutifully doubled the dimensions of the altar, with no perceptible effect on the plague. Their belated realization that the oracle had meant doubling the *volume*, and not the *dimensions*, of the altar is supposed to have led to the problem of doubling the cube.

It seems to have been *Hippocrates* (*c*.450 B.C.) who made the first real progress in the problem of doubling the cube. He recognized that the problem could be reduced to the construction of a pair of *mean proportionals* between two given line segments of arbitrary given lengths a, b, that is, the construction of two line segments of lengths x, y standing in the ratios $a : x = x : y = y : b$. For if $b = 2a$, then, upon eliminating y, we obtain $x^3 = 2a^3$, so that x is the length of the side of a cube of double the volume of one of side a.

While the problem of doubling the cube continued to elude solution using only Euclidean tools, Hippocrates's breakthrough made it possible for his successors to elaborate constructions which "solved" the problem in a manner *not strictly Euclidean*. Two such "solutions" were furnished by *Menaechmus*, a pupil of Eudoxus, around 350 B.C. His arguments are based on the idea of a *conic section*, a concept he is reputed to have introduced for the express purpose of solving the problem. Menaechmus defined a conic section to be a curve bounding a section of a right circular cone cut off by a plane perpendicular to the cone's surface. According as the vertex angle of the cone is less than, equal to, or greater than, a right angle, the curve obtained is an *ellipse*, a *parabola*, or a *hyperbola*. (These terms are later coinages: Menaechmus himself would have used the descriptive words *oxytome*, "sharp cut", *orthotome*, "right cut", and *amblytome*, "dull cut".) Menaechmus showed that a solution to the problem of constructing two mean proportionals, and hence to that of doubling the cube, could be achieved by constructing a pair of parabolas, or by constructing a parabola and a hyperbola. In modern terms, his argument went as follows: to find two mean proportionals between a, b, we seek x, y so that $a/x = x/y = y/b$, that is, to satisfy the equations

$$x^2 = ay, \quad y^2 = bx, \quad xy = ab.$$

The required values can thus be obtained from the points of intersection of the parabolas $x^2 = ay$, $y^2 = bx$:

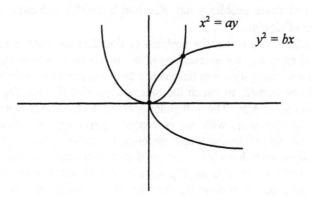

or of the parabola $x^2 = ay$ with the hyperbola $xy = ab$:

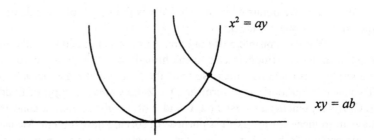

Another such "solution" to the problem of doubling the cube was devised by *Apollonius of Perga* (*c.*262–190 B.C.), an outstanding mathematician of antiquity. Draw a rectangle *OACB*, where *OB* is twice the length of *OA*, and then a circle, concentric with *OACB*, cutting *OA* and *OB* produced in *A'* and *B'* such that *A'*, *C*, *B'* are collinear. It can then be shown that $(BB')^3 = 2.(OA)^3$. Actually, it is impossible to construct the specified circle with Euclidean tools, but Apollonius did manage to formulate a mechanical way of describing it.

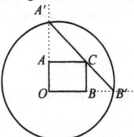

To his contemporaries it was Apollonius, and not Euclid, who was known as "the Great Geometer," a title which was deservedly bestowed on him in recognition of his authorship of the masterly treatise, the *Conics*, one of only two of his works which have come down to us. In this work Apollonius assigns the names *ellipse, parabola,*

and *hyperbola* to the conic sections, terms which had originally been employed, probably by the Pythagoreans, in the areal solution of quadratic equations. This method had involved the placing of the base of a given rectangle along a line segment so that one end of each coincided. According as the other end of the rectangle's base fell short of, coincided with, or fell beyond the other end of the line segment, it was said that one had *ellipsis* ("deficiency"), *parabole* ("thrown alongside") or *hyperbole* ("thrown beyond"). Using the modern representation of conic sections in coordinate geometry, it is not difficult to see why Apollonius applied these terms in this new context. The equation of a parabola with its vertex at the origin is $y^2 = \ell x$, so that, at any point on it, the square on the ordinate y is precisely equal to the rectangle on the abscissa x and the parameter ℓ. The equations of the hyperbola and the ellipse, again with a vertex as origin, are $(x \pm a)^2/a^2 \mp y^2/b^2 = 1$, or $y^2 = \ell x \pm b^2 x^2/a^2$, where $\ell = 2b^2/a$. Thus, for the ellipse, $y^2 < \ell x$ and for the hyperbola $y^2 > \ell x$. So for the ellipse, hyperbola and parabola, we have $y^2/x <, =,$ or $> \ell$, respectively, that is, the base of the rectangle with area y^2 and width x falls short of, coincides with, or exceeds the line segment of length ℓ.

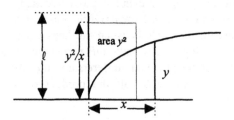

In the *Conics* Apollonius shows that all three varieties of conic section can be obtained from a single cone by varying the inclination of the cutting plane, thus establishing the important fact that all three are *projections* of each other, and showing also that the circle is a special type of ellipse (obtained when the cutting plane is parallel to the base of the cone). Apollonius also shows that sectioning an oblique or scalene cone yields the same curves as does a right cone. Finally he took the significant step of replacing the single cone by the double cone obtained by allowing a straight line, with one point fixed, to move around the circumference of a given circle. In this way he revealed that the hyperbola has two branches, corresponding to sections of both sheets of the double cone.

Of all the famous problems of mathematics, that of *squaring the circle* has exerted the greatest fascination through the ages[4]. This problem also eluded solution with Euclidean tools, but, again, Greek mathematicians produced "solutions" which, although not Euclidean, possessed great ingenuity and elegance. One such solution is due to *Archimedes* (287–212 B.C.), regarded as the greatest mathematician of antiquity, and one of the greatest of all time. His solution of the problem was achieved by means of a special curve known as the *spiral of Archimedes*. We may define the spiral in

---

[4] This problem was apparently sufficiently familiar to the Athenian public of the 4th century B.C. for Aristophanes (c.450–c.385 B.C.) to allude to it in one of his comedies, *The Birds*.

kinematic terms as the locus of a point P moving uniformly along a ray which, in turn, is uniformly rotating about its end point in a plane. (More concretely, the spiral is the

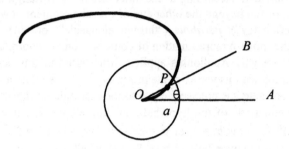

path traced out by a bug crawling along a rotating rod.) If we write r for the distance of P from the end point of the ray when it makes an angle θ with its initial position (i.e., when P is at the end point O), then r and θ are related by an equation of the form $r = a\theta$, where a is some constant. Now draw the circle with centre at O and radius a. Then OP and the arc of this circle between the lines OA and OP are equal, since each is given by $a\theta$. Accordingly, if we take OP perpendicular to OA, then OP will have a length equal to one quarter of the circumference of the circle. Since—as the Greeks knew—the area K of the circle is half the product of its radius and its circumference, we get

$$K = (a/2).(4OP) = (2a).OP.$$

In other words, the area of the circle is the same as the area of the rectangle whose sides are the diameter of the circle and the radius vector of the spiral which is perpendicular to OA. Since it is an easy Euclidean construction to produce the side of a square equal in area to that of a given rectangle, we get a solution to the problem of squaring the circle.

Archimedes was also a master of the method of exhaustion, and used it to give rigorous proofs of the formulas for areas and volumes of quite complex figures, for example parabolic segments, and ellipsoids and paraboloids of revolution. Observing that the method of exhaustion is essentially a tool for *proof* but not for *discovery*, one may ask how Archimedes actually *found* his formulas. This only became known with the discovery in Constantinople in 1906 of a copy of Archimedes' long-lost treatise *The Method*, addressed to the mathematician *Eratosthenes* (*c.*230 B.C., famous for his remarkably accurate estimate of the diameter of the earth). The manuscript was found in what is known as a palimpsest: it had been copied onto parchment in the tenth century which was later—in the thirteenth century—washed off and reused for inscribing a religious text. This, it may be remarked, provides an excellent illustration of the intellectual priorities of the times! Fortunately, most of the original text was restorable from beneath the later one.

In *The Method*, Archimedes reveals that in discovering some of his remarkable area and volume formulas he employed ideas from the physical science of statics which was itself another product of his fertile mind. The fundamental idea was this: to find the area or volume of a given figure, (mentally) cut it up into a very large number of parallel thin strips or layers, and (again mentally) hang these pieces at one end of a given lever in such a way as to be in equilibrium with a figure whose content and center of gravity are known. We illustrate this idea by using it to obtain the formula for the *volume of a spherical segment*. In the derivation we shall make use of the *law of the lever*, which states that two bodies on a lever balance when their weights are inversely proportional to their distances from the fulcrum. Accordingly two bodies will balance if they have equal moments about the fulcrum, where the *moment* of a body about a point is defined to be the product of the mass of the body with the distance of its centre of gravity from the point.

Thus consider a segment of height h of a sphere of radius r, positioned with its polar diameter along the x-axis with its north pole N at the origin. By rotating the rectangle $NABS$ and the triangle NCS about the x-axis, we obtain a cylinder and a cone. We assume that sphere, cone and cylinder are homogeneous solids of unit density. Now cut from these three solids thin vertical slices (assuming that these are, approximately, flat cylinders) at distance x from N and thickness ε. The volumes of these slices are, approximately,[5]

$$\text{sphere: } \pi x(2r-x)\varepsilon$$
$$\text{cylinder: } \pi r^2$$
$$\text{cone: } \pi x^2 \varepsilon.$$

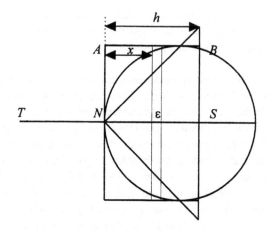

Now let T be a point on the x-axis at distance r from the origin in the opposite direction from S. If the slices of the sphere and cone are suspended at T, then their combined moment about N is

$$[\pi x(2r - x)\varepsilon + \pi x^2\varepsilon].r = 2\pi r^2 x\varepsilon.$$

This is clearly twice the moment about N of the slice cut from the cylinder when that slice is left where it is. In other words, the slices of the sphere and cone placed at T exactly balance —with fulcrum placed at N— twice the slice of the cylinder left where it is. Adding all these slices together, the results will still balance, so that, about N,

moment of mass of spherical segment + mass of cone concentrated at N
= 2.moment of cylinder.

Assuming it is known that the volume of a cone is ⅓ base area × height, and that of a cylinder is base area × height (facts both familiar to Archimedes), we get

$$r. \text{[volume of spherical segment} + \pi h^3/3] = 2.h/2.\pi r^2 h,$$

from which we deduce that the volume of the spherical segment is

$$\pi r h^2 - \pi h^3/3 = \pi h^2(3r - h)/3.$$

From this it is easy to deduce Proposition 2 of *The Method*, which Archimedes states in the following way:

Any segment of a sphere has to the cone with the same base and height the ratio which the sum of the radius of the sphere and the height of the complementary segment has to the height of the complementary segment.

Archimedes wrote many masterly mathematical treatises. Those that have come down to us include *On the Sphere and Circle, The Measurement of the Circle, On Conoids and Spheroids, On Spirals, On the Quadrature of the Parabola, On the Equilibrium of Planes, On Floating Bodies,* and *The Sand-Reckoner.* In the first of these Archimedes obtains, by strictly geometric means, the volumes and surface areas of spherical segments, cones, and cylinders, establishing in particular that the ratio of the volume of a right circular cylinder to that of an inscribed sphere is 3:2 (a representation of which was inscribed on his tomb), and that the area of a sphere is exactly 4 times that of a great circle. In *The Measurement of a Circle* he shows, using inscribed and circumscribed polygons of 96 sides, that the circumference of a circle is bounded between 22/7 and 223/71 times its diameter. *On Conoids and Spheroids* contains an investigation of the properties of solids obtained by the revolution of conic sections about their axes, the main results being the comparison of the volume of a segment of such a solid cut off by a plane with that of a cone having the same base and axis: for a paraboloid of revolution he shows that the ratio is 3:2. Archimedes also determines the area of an ellipse: as he puts the matter, "the areas of ellipses are as to the rectangles under their axes." Most of the proofs in this work employ the method of exhaustion. *On Spirals*, regarded by Archimedes' successors as his most recondite work, contains, in addition to the argument for squaring the circle described above, a number of propositions concerning the areas swept out by the radius vector of the spiral. In *On the Quadrature of the Parabola*, said to have been Archimedes' most popular work, he shows that the area of a parabolic segment is 4/3 times that of the inscribed triangle. This was the first quadrature of a conic section. *On the Equilibrium of Planes* contains Archimedes' pioneering work on statics. Here he derives the law of the lever and locates the centres of gravity of various figures, including triangles, trapezoids, and parabolic segments. In *On Floating Bodies* Archimedes starts from a simple postulate concerning fluid pressure and proceeds to create the science of hydrostatics. He derives the now familiar principle of buoyancy that a floating solid displaces its own weight of fluid—a discovery which is, famously, reputed to have induced him to leap from the bath and run naked through the streets shouting *Eureka!*, "I have found it!". In this work he also obtains complete determinations of the positions of equilibrium of floating paraboloids of revolution, a problem possibly arising in connection with the design of ship's hulls.

The Sand-Reckoner occupies a special place in Archimedes' output, for here he is concerned not with a geometric problem, but with a notational one. In this short work he shows that any finite set A, however large, can always be enumerated in the strong sense that a number exceeding that of the number of elements of A can always be *named*. Taking A to be a set of grains of sand filling the entire universe, he elaborates a system of notation, based on powers of a *myriad myriads*, i.e., 10^8, in which he is able to express an upper bound for the number of elements of A. By "universe" Archimedes meant the (finite) cosmos as conceived by *Aristarchus of Samos* (c. 310 – 230 B.C.), famous for his calculations of the distances of the sun and moon from the earth.[6] Archimedes calculated that the number of grains of sand required to fill a sphere the size of Aristarchus's universe is less than a number we would write as 10^{63}. Archimedes actually extended his notation to embrace numbers up to 10 to the power 8×10^{15}.

*

The idea of a *deductive* or *demonstrative* science, with mathematics as a prime example, seems to be original with the Greeks. But, one may ask, what prompted them to transform mathematics into such a science, an idea which appears to have been unknown to their predecessors? The pre-Greek mathematical documents that have come down to us are all concerned with practical or empirical questions; in none of them do we find a *theorem* or a *demonstration*, a *definition*, a *postulate*, or an *axiom*. As far as is known, these concepts originate with Greek mathematicians. But why were

---

[6] Aristarchus had also put forward a heliocentric theory of the solar system, but this was not accepted by most of the leading Greek astronomers, Archimedes included.

they introduced? Why, in other words, did Greek mathematicians come to put more trust in theoretical demonstration than in what could be verified by practice alone?

This turning-point in Greek thought has been attributed to several sources: the advanced sociopolitical development of the Greek city-states; the effort to isolate the general principles underlying the diverse methods for solving mathematical problems that had been inherited from the Egyptians and Babylonians; the influence of the art of disputation. Perhaps each of these played some role. An intriguing theory has been put forward by the historian of mathematics *Arpad Szabó*, who believes that Greek mathematics became a deductive science through the introduction by the Pythagoreans of the technique of *indirect demonstration*, an idea which he thinks was inherited from the *Eleatic* school of philosophy (fl. 550–400 B.C.), whose members included Parmenides and Zeno. The principal doctrines of the Eleatics were developed in opposition to the physical theories of the early Greek materialist philosophers, such as Thales and *Anaximander* (fl. *c.*560 B.C.) who explained all existence in terms of primary substance, and also to the theory of *Heraclitus* (fl. *c.*500 B.C.) that existence is perpetual flux. As against these theories the Eleatics maintained that the world is an unchanging unity and that the impression to the contrary arising through the senses is an illusion. In arguing for these claims they made constant use of the technique of indirect demonstration or *reductio ad absurdum*, that is, establishing the truth of a proposition by demonstrating the absurdity of its contrary[7].

Szabó conjectures that the Pythagoreans were moved to adopt the indirect form of reasoning after all efforts to generate a common measure for the side and diagonal of a square had failed. Beginning to suspect that such a common measure does not exist, they saw that the only possible way in which this nonexistence could be conclusively established was by means of an indirect argument in the Eleatic manner. Thus they started by assuming the contrary possibility, that is, the commensurability of the two segments. From this, as we have seen above, they derived the contradiction that a number can be both even and odd. In this way they demonstrated the absurdity of the original supposition, so conclusively establishing that no common measure could exist.

Szabó believes that initially Greek mathematics was, like its near-eastern predecessors, of a primarily illustrative character, its arguments built on concrete visualization, making essential use of figures. But, with the assimilation of the abstract and anti-illustrative Eleatic modes of reasoning, Greek mathematicians could no longer appeal to visual intuition to justify their arguments and were instead forced to make explicit the principles and assumptions on which their reasoning rested. It is this fact, Szabó suggests, which ultimately led the Greeks to cast their mathematics in deductive form.

---

[7] The most famous Eleatic arguments of this kind are *Zeno's paradoxes* which purport to demonstrate the impossibility of motion: see Ch. 10

CHAPTER 3

THE DEVELOPMENT OF THE NUMBER CONCEPT

THE CONCEPT OF NUMBER HAS undergone a long evolution, and today there are several types of "number". Let us see how this has come about.

The simplest numbers are, of course, the *natural or whole numbers* 1, 2, 3,... associated with the process of *counting*. The basic operations of *addition* and *multiplication* of natural numbers reflect simple methods of *combining* assemblages of objects. Thus, for example, in calculating 2 + 3, one starts with 2 = ** objects and juxtapose 3 = *** objects; counting the result yields a set of ** *** = 5 objects. The same number 5 is arrived at by the reverse process of starting with 3 objects and juxtaposing 2 objects: ** *** = *** **. Since this discussion is independent of the particular numbers 2, 3 concerned, we are led to adopt the general rule known as the *commutative law of addition:* $m + n = n + m$ for any natural numbers m, n. Similar considerations lead to the recognition of the truth of the *associative law of addition:* $(m + n) + p = m + (n + p)$ for any natural numbers m, n, p. *Multiplication* now arises from repeated addition: 2×3 (also written 2.3), for example, corresponds to the combination of 2 sets of 3 objects. This contains altogether 6 objects, so $2 \times 3 = 6$. Clearly this combination

may also be obtained by combining 3 sets of 2 objects, so $3 \times 2 = 6$. Thus we are led to adopt the general rule known as the *commutative law of multiplication:* $m \times n = n \times m$ for any natural numbers m, n. Similar considerations lead to the recognition of the truth of the *associative law of multiplication:* $m \times (n \times p) = (m \times n) \times p$ for any natural numbers m, n, p. We also recognize an important link between addition and multiplication—the *distributive law*—which arises in the following way. The expression $2 \times (3 + 4)$, for example, is presented as

We may regard this collection as being composed of two 3s and two 4s, i.e. as

Thus $2 \times (3 + 4) = 2 \times 3 + 2 \times 4$, and we are led to the general assertion $m \times (n + p) = (m \times n) + (m \times p)$ for any natural numbers m, n, p. This is the *distributive law* for multiplication over addition.

The most important feature of the system of natural numbers is that it satisfies the *Principle of Mathematical Induction*, which may be stated as follows. *For any property* P, *if* 1 *has* P *and, whenever a natural number* n *has* P, *so does* n + 1, *then every natural number has* P. For if the property P satisfies the premise, then 1 has P; from this it follows that 2 has P, from this in turn that 3 has P, and so on for all n. This principle is implicit in Euclid's proof of the infinitude of the set of prime numbers: for he shows that, if there are n primes, there must be $n + 1$ primes; and since there is 1 prime, it follows that there are n primes for every n, that is, there are infinitely many primes.

The principle was recognized explicitly by the Italian mathematician *Francesco Maurolico* (1494 – 1575) in his *Arithmetica* of 1575. He used it to prove, for example, the fact—known to the Pythagoreans—that

$$1 + 3 + 5 + \ldots + (2n - 1) = n^2. \tag{1}$$

This is easily established by mathematical induction. Write $P(n)$ for the property of n asserted by equation (1). Then clearly 1 has P. Now suppose that n has P, i.e., that (1) holds. Then

$$1 + 3 + 5 + \ldots + (2n - 1) + (2n + 1) = n^2 + 2n + 1 = (n + 1)^2,$$

so that $n + 1$ also has P. From the principle of mathematical induction we conclude that every number has P; in other words, (1) holds for every n.

Chronologically the first *enlargement* of the system of natural numbers occurred with the adjoining of *fractions*. These owe their invention to the transition from *counting* to *measuring*. Any process of measuring starts with a domain of *similar magnitudes*, such as, for example, the segments **a**, **b** on a straight line. In this case we have, first, a relation **a** = **b** of *equality* or *congruence* of segments (**a** and **b** are said to be *congruent* if their endpoints can be brought into coincidence) and secondly, an operation + of *juxtaposition* or *addition* applicable to any pair **a**, **b** of segments producing a segment **a** + **b** called their *sum*. By *iterating* the operation of addition on a single segment **a** we obtain for example the segment 4**a** as the sum **a** + **a** + **a** + **a** with 4 terms **a**; in general, for any natural number n, the nth *iterate* n**a** is defined to be **a** + **a** + ... + **a** with n terms **a**. In this way each natural number n comes to *symbolize an operation*, namely, the operation "repeat addition n times."

The operation of iterated addition on line segments has an *inverse* called *division*: given a segment **a** and a natural number n, there is a unique[1] segment **x** such that $nx = a$; it is denoted by a/n and is the segment obtained by *dividing a into n equal parts*. Note that, in claiming that division can always be carried out, no matter how large n may be, we are implicitly assuming that our magnitudes—in this case, line segments—are *continuous*, that is, have no "smallest" parts which are incapable of being further divided.

The operations of iterated addition and division can be *combined*: thus, for natural numbers m,n and a segment **a** we define ma/n to be the unique segment **x** for which $ma = nx$. The *fraction* m/n serves then to denote the *composite* operation "repeat addition m times and divide the result into n equal parts." Two fractions are accordingly deemed to be *equal* if the operations they denote are the same, that is, if they both lead to the same result no matter to what segment **a** they may be applied: it is then readily shown that

$$m/n = p/q \text{ if and only if}^2 \ mq = np. \tag{2}$$

Multiplication of fractions is performed by *composition*, that is, by carrying out successively the operations denoted by them: thus

$$(m/n).(p/q) = mp/nq. \tag{3}$$

Addition of fractions is not defined quite so readily, but may be performed by noting that, for given m, n, p, q and an arbitrary segment **x**, we have the identity

$$mx/n + px/q = [(mq + np)/nq]x,$$

so that we may define

$$m/n + p/q = (mq + np)/nq. \tag{4}$$

It should be clear that the cogency of this discussion in no way depends upon the specific nature of the magnitudes under consideration; it is only necessary that the operations of (iterated) addition and division be defined on them. Thus we do not need to define special fractions for each domain of magnitudes: just as one system of *natural numbers* is intended to serve for *counting all collections of objects*, so, likewise, one system of *fractions* is intended to serve for *measuring all domains of magnitudes*. It follows that each pair of natural numbers may be held to determine a unique fraction, and conversely. The definitions of equality, addition and multiplication for fractions conceived as pairs of numbers in this way are then given by (2), (3) and (4) above. It is readily verified that the commutative, associative and distributive laws continue to hold in the system of fractions. Each natural number n may be identified with the fraction

---

[1] By *unique* here we mean *unique up to congruence*, that is, all such segments are congruent.

[2] Here we follow the usual convention and omit the dot in multiplication.

$n/1$, and so regarded as a special kind of fraction (i.e., having unit denominator). Thus the system of fractions becomes an *enlargement* of the system of natural numbers.

Fractions were invented essentially in order to symbolize the operation inverse to multiplication. Another important development was the introduction of *negative numbers,* which arise as the result of conceiving of an operation—*subtraction*—*inverse to addition.* This is not quite so easily done since, while fractional magnitudes are encountered in everyday life, it is not immediately clear what meaning is to be assigned to the idea of a "negative" magnitude, and as a result negative numbers were not fully accepted until comparatively late in the development of mathematics. Ultimately it came to be seen that negative numbers arise naturally in connection with the measurement of what we may call *oriented magnitudes,* e.g., financial profits and losses, or *directed line segments.* We use these latter to illustrate the idea.

Accordingly, we now suppose that each line segment **a** is assigned a *direction* or *orientation* (to the right or left). *Equality* **a** = **b** of directed line segments **a,b** is then taken to mean that they are not only congruent but also have the *same direction.* The *sum* **a** + **b** is the directed line segment obtained as follows: if **a** and **b** have the *same* orientation, then **a** + **b** is the line segment, with that same orientation, obtained by juxtaposing them. If **a** and **b** have *opposite* orientations, but *different* lengths, then one, **a** say, is the greater; we then define **a** + **b** to be the line segment with orientation that of **a** obtained by removing a segment of length **b** from **a**. Finally, if **a** and **b** have opposite orientations but *identical* lengths, we define **a** + **b** to be a *lengthless* line segment, that is, a segment **0** whose sum with any line segment **a** is just **a**.

Once the sum of directed line segments has been introduced we can define for each natural number n and each **a** the iterate n**a** as before. This enables us to define the symbol $-n$ by stipulating that $-n$**a** = $(n$**a**$)^*$, where, for each segment **x**, **x*** denotes the segment obtained by *reversing* the orientation of **x**. The symbol $-n$ thus signifies the operation "repeat addition n times and reverse orientation." Clearly, for any natural number n and any segment **a**, we have

$$n\mathbf{a} + (-n\mathbf{a}) = \mathbf{0}.$$

We also introduce the symbol 0 by defining it to symbolize the operation which, upon application to any segment **x**, reduces it to the lengthless segment **0**, i.e.,

$$0\mathbf{x} = \mathbf{0}.$$

We have thus obtained an enlarged system of symbols ...,-3, -2, 1, 0, 1, 2, 3, ... each of which denotes a certain *operation* on directed line segments. *Addition* and *subtraction* is defined on these symbols by stipulating that, for each pair p, q, $p + q$, $p - q$ are to be the unique operations such that, for all **x**,

$$(p + q)\mathbf{x} = p\mathbf{x} + q\mathbf{x} \qquad (p - q)\mathbf{x} = p\mathbf{x} + (q\mathbf{x})^*.$$

Multiplication is defined, as for fractions, by composing the corresponding operations, i.e. by stipulating that, for any **x**,

$$(p.q)\mathbf{x} = p(q\mathbf{x}).$$

Clearly the precise nature of the magnitudes we have employed in our discussion is irrelevant, so that—as in the case of fractions—the whole system ...,−3, −2, −1, 0, 1, 2, 3,... may be regarded as autonomous. It is called the *set of integers*: 1, 2, 3,... are called the *positive integers* (or natural numbers),...−3, −2, −1 the *negative integers* and 0 the *number zero*.

It is easy to verify that the operation of subtraction on integers as defined above is the *inverse* of addition: i.e., for any integers p, q, $p − q$ is the unique integer x satisfying $q + x = p$. So the set of integers is, as intended, an enlargement of the set of natural numbers on which the operation of addition *has an inverse*. It is also readily verified that the *commutative, associative,* and *distributive laws* continue to hold in the set of integers.

We have thus shown how to enlarge the set of natural numbers to a system in which multiplication has an inverse—the fractions—and also to one in which addition has an inverse—the integers. If we perform the construction of fractions, only this time starting with the *integers* instead of the natural numbers (in other words, using directed magnitudes in place of neutral ones), we obtain an enlargement of the system of natural numbers in which *both* addition *and* multiplication have inverses (only excepting multiplication by 0). This enlargement is called the system of *rational numbers*. Any rational number may be represented in the form of a fraction p/n where p is an integer and n is a natural number. The various laws satisfied by the rational numbers may be summarized as follows: for all rational numbers x,y,z,

$$x + y = y + x, \quad (x + y) + z = x + (y + z), \quad 0 + x = x + 0 = x, \quad x − x = 0$$

$$x.y = y.x, \quad x.(y.z) = (x.y).z, \quad x.1 = 1.x = x, \quad x.(1/x) = 1 \text{ (provided } x \neq 0)$$

$$x.(y + z) = x.y + x.z.$$

A system of objects or symbols with two operations + and . defined on it, which contains two distinguished objects 0 and 1, and which satisfies the above conditions is what mathematicians call a *field*. We may sum up the fundamental character of the system of rational numbers by saying that it *constitutes a field*. The field of rational numbers is denoted by **Q** (from German *quotient*). Note also that the system of positive and negative integers satisfies all the field conditions with the exception of that involving the reciprocal $1/x$: such a system is called a *ring*. The ring of integers is denoted by **Z** (from German *zahl*, "number").

It is helpful to think of the rational numbers as *points on a line*, in which the positive rationals (i.e. those of the form m/n with m and n positive integers) appear to the right of 0 and the negative rationals (i.e those of the form p/q with p negative and q positive) to the left of 0. This representation displays the *ordering* of the rational

$$-3 \;\; -2 \;\; -1 \;\; -½ \quad 0 \quad ½ \quad 1 \quad 3/2 \quad 2 \quad 3$$

numbers: thus $p/m < q/n$ means that the integer pn is to the left of the integer qm. This ordering turns the rational numbers into what is known as an *ordered field*.

Regarding the rationals as points on a line enables us to divide the line as finely as we please. For example, to represent all rationals of the form $m/10^9$ as points on the line, we divide the interval $(0, 1)$ of the line between 0 and 1 into a billion equal pieces; similarly for all other intervals $(1, 2)$, $(2, 3)$,... and the points of subdivision then correspond to fractions of the form $m/10^9$. Since the denominators of these fractions can be made arbitrarily large, 10^{10}, 10^{100}, or whatever, thereby producing subdivisions of unlimited fineness, it would be natural to suppose that in this way one would capture as points of subdivision *all* the points on the line—in other words, that every point is represented by a rational number. Now it is certainly true that rational numbers suffice for all practical purposes of measuring. Moreover, the rational points are *dense* on the line in the sense that they may be found in any interval (a, b), however small, with rational endpoints $a < b$. To verify this, observe that the rational number $(a + b)/2$ lies between a and b. It may actually be inferred from this fact that each such interval contains *infinitely* many rational points. For if the interval (a, b) contained only finitely many, n say, then we could mark them off as shown, and then any interval between two adjacent points would be free of rational points, contradicting what we have already established.

$$a \hspace{6cm} b$$

All this seems to lend support to the idea that every point on the line is represented by a rational number. The Pythagorean discovery of the incommensurability of the side and diagonal of a square shows, however, that this idea is incorrect. If we take two perpendicular lines OP and PQ of length 1 and use compasses to mark out on the line ℓ a line OR of the same length as OQ then the incommensurability of OP and OQ, and hence also OR, just means that R is not a

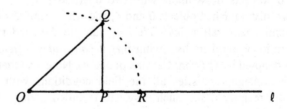

rational point. In fact, if we designate by r the "number" associated with the point R, then we have, by the Pythagorean theorem, $r^2 = 1^2 + 1^2 = 2$, so that, replacing r by the customary symbol $\sqrt{2}$, we conclude that *no rational number is equal to $\sqrt{2}$, or that $\sqrt{2}$ is irrational.*[3]

It can be shown, by arguments similar to that establishing the irrationality of $\sqrt{2}$, that other numbers formed by root extraction, such as $\sqrt{7}$, $\sqrt[3]{2}$ (these are solutions to the equations $x^2 = 7$, $x^3 = 2$) are also irrational. Another number that turns out to be irrational (although this is more difficult to prove) is π, the length of the circumference of a circle of unit diameter.

The fact that not every point on the line corresponds to a rational number means that, if we want to maintain the correspondence between points on a line and "numbers", the system of rational numbers will have to be *enlarged still further*. This leads to the system of *real numbers*. In the last century several methods of constructing this system were devised: here we outline the most straightforward one, which is an extension of the *decimal representation* of rational numbers. The first thing to observe is that a decimal fraction represents a rational number precisely when it is *periodic*, i.e., displays a repeating pattern indefinitely after a certain stage (we regard finite decimals as special cases of periodic ones). For example,

$$6\ 5/8 = 6.625, \quad 3/7 = 0.428571\ 428571\ .....$$

The fact is easy to establish: consider the usual long division of N into M which gives the decimal form of M/N. The only possible remainders at each stage of the division are $1, 2,..., N - 1$, so that there are just $N - 1$ possibilities: if any remainder is 0 the process stops and a finite decimal results. It follows that, after at most N divisions, one of the remainders *must* repeat. Therefore, since each remainder is uniquely determined by its predecessor, the subsequent remainders, and so also the decimal representation itself, must repeat. (Clearly the "repeating block" can have at most $N - 1$ digits.) Conversely, any periodic decimal represents a rational number. For example, consider

$$r = 0.90909090...$$

This may be converted into a fraction by writing

$$100r = 90.909090...$$

and subtracting: we obtain $99r = 90$, so that $r = 90/99 = 10/11$. This procedure may be applied to any periodic decimal: if the repeating block contains m digits we multiply by 10^m and subtract just as we have done above.

It follows from this that any *nonperiodic* decimal must represent an *irrational* number; it is easy to furnish examples of these, for instance the following decimal containing an increasing number of zeros:

$$0.101001000100001000001....$$

---

[3] The term "irrational" here is to be understood in the sense of "that which cannot be expressed as a ratio" as opposed to its more usual (but related) meaning "contrary to reason".

Surprisingly, perhaps, no simple rule of this kind exists for constructing the decimal representation of familiar irrational numbers such as $\sqrt{2}$.

The *real numbers* are now defined to be the set of all finite or infinite (positive or negative) decimals: considered geometrically the real numbers constitute the *geometric continuum* or *real line*. The operations of addition and multiplication can be naturally extended to the real numbers so that, like the rational numbers, they constitute a *field*, which we shall denote by the symbol **R**. So the real numbers resemble the rational numbers insofar as they are subject to the same operational laws. On the other hand, if we regard the integers as the basic ingredients from which the other numbers are constructed, then our discussion shows that, while each rational number can be defined in terms of just two integers, in general a real number requires *infinitely many* integers to define it. The fact that infinite processes play an essential role in the construction of the real numbers places them in sharp contrast with the rationals.

In any case the simple definition of real numbers as infinite decimals we have given is not entirely satisfactory since, for one thing, there is no compelling mathematical reason to choose the number 10 as a base for them. We also recall that the real numbers were introduced in order to correspond exactly to points on a line: but how do we know that every such point corresponds to a real number thus defined as a decimal? To establish this it is necessary to show that there are no "gaps" in our set of real numbers, and to define with precision what is to be understood by this assertion. This was carried out in the latter half of the nineteenth century and resulted in the modern theory of real numbers.

The last extension of the concept of number that we shall consider here —the *complex numbers*—arose, not for geometric reasons, but as the result of attempting to formulate solutions to certain algebraic equations. We have seen that, in order to solve a quadratic equation such as $x^2 = 2$ we need to introduce irrational numbers, and that the rational and irrational numbers together comprise the real numbers. Thinking of the real numbers as points on a line, they are ordered from left to right, with the negative real numbers to the left of zero and the positive ones to the right. Now since the square of any real number, positive or negative, is always positive (or zero) one sees immediately that there can be no real number whose square is *negative*; in particular, no real number x exists which satisfies $x^2 = -1$. That is, even the simple quadratic equation $x^2 + 1 = 0$ *cannot be solved in the system of real numbers.* In order to be able to solve such equations we are obliged once again to enlarge our number system. This can be done by formally introducing a new symbol i (called the *imaginary*[4] *unit*) which is postulated to satisfy the equation

$$i^2 = -1.$$

---

[4] The term "imaginary"—first used in this connection by Descartes—reflects the fact that seventeenth-century mathematicians considered square roots of negative numbers (such as $i = \sqrt{-1}$) to be fictitious, the mere product of imagination. In this respect it is to be contrasted with the term "irrational": *q.v.* footnote 3.

We shall suppose that we can form multiples bi of i by any real number b, and sums $a + bi$ for any real number a. A sum of the form $a + bi = z$ is called a *complex number*: a is called the *real* part and b the *imaginary* part, of z. We assume that complex numbers can be added and multiplied in the same way as real numbers, and, in particular, that these operations satisfy the same laws, so that the complex numbers constitute a *field*, which we denote by the symbol **C**. Calculations with complex numbers can then be performed as with real numbers, replacing i^2, wherever it occurs, by -1. Thus, for example, we can compute the product $(2 + 3i)(4 + 5i)$ as follows:

$$(2 + 3i)(4 + 5i) = 8 + 10i + 12i + 15i^2$$
$$= (8 - 15) + (10 + 12)i$$
$$= -7 + 22i.$$

Each real number a may be identified with the complex number $a + 0i$, so that the field of complex numbers may be regarded as an *enlargement* of the field of real numbers (and thus ultimately as an enlargement of the set of natural numbers).

We observe that two complex numbers $z = a + bi$ and $w = c + di$ are equal if and only if $a = c$ and $b = d$: this fact is known as the *principle of equating real and imaginary parts*.

Although the field of complex numbers was introduced merely to provide solutions to quadratic equations of the form $x^2 + a = 0$ (which has solutions $i\sqrt{a}$, $-i\sqrt{a}$ there), much more has actually been gained. In fact, *every* quadratic equation $ax^2 + bx + c = 0$ (a, b c real and $a \neq 0$) has exactly two complex roots, given by the familiar formula

$$x = [-b \pm \sqrt{(b^2 - 4ac)}] / 2a.$$

These roots are real if $b^2 - 4ac \geq 0$ and complex otherwise. In general, it can be shown that *any* algebraic equation—with real *or* complex coefficients—can be solved in the field of complex numbers. This result, known as the *Fundamental Theorem of Algebra*, shows that, with the construction of the field of complex numbers, the task of extending the domain of real numbers so as to enable all algebraic equations to be solved has been completed.

Unlike real numbers, complex numbers cannot be represented as points on a line since there is no simple order relation on them. Nevertheless, at the beginning of the nineteenth century they were furnished—independently, and more or less simultaneously, by *Caspar Wessel* (1745–1818), *Jean-Robert Argand* (1768–1822) and *Karl Friedrich Gauss* (1777–1855)—with a simple geometric interpretation as points in a *plane*. In this interpretation, the complex number $a + bi$ is represented as the point z in the plane with rectangular coordinates (a, b). In this way any complex number is

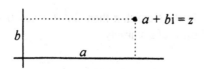

uniquely correlated with a point in what is known as the *complex plane* (or *Argand diagram*). Addition and multiplication of complex numbers then admit simple geometric interpretations. In the case of addition, we regard $a + bi$ as a displacement (or *vector*) from the origin $(0, 0)$ to (a, b). Thus the sum

$$(a + bi) + (c + di) = (a + c) + (b + d)i$$

is the point in the complex plane obtained by the vector (parallelogram) addition law:

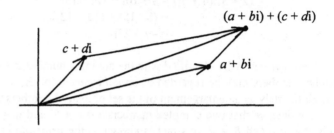

To explain the geometric interpretation of multiplication of complex numbers we shall need to introduce their *trigonometric representation*. Referring to the figure immediately below, if we denote the distance of the point z from the origin O by r, then by the Pythagorean theorem $r = \sqrt{(a^2 + b^2)}$. This real number is called the *modulus* of z, and is written $|z|$. Note that the distance between the points represented by the complex

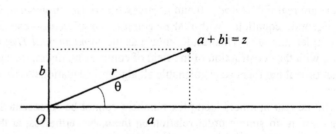

numbers z and z' is then $|z - z'|$. The angle θ that the line Oz makes with the positive x-axis is called the *argument* or *amplitude* of z. In trigonometry the *sine* and *cosine* of the angle θ are defined by

$$\sin \theta = b/r \quad \cos \theta = a/r,$$

so that $a = r\cos \theta$, $b = r\sin \theta$. (Note that then $\sin^2 \theta + \cos^2 \theta = (a^2 + b^2)/r^2 = 1$.) This yields the trigonometrical form of z, namely

$$z = a + bi = r(\cos \theta + i \sin \theta). \tag{5}$$

Thus, for example, if $z = 1 + i$, then $r = \sqrt{2}$, $\theta = 45°$, so that

$$1 + i = \sqrt{2}(\cos 45° + i \sin 45°).$$

The representation (5) yields a simple expression for the reciprocal of a complex number. In fact, if z is given as in (5), we find that, assuming $r \neq 0$,

$$1/z = 1/r.(\cos \theta - i \sin \theta).$$

This follows from the calculation

$$z.1/z = r.\ 1/r.\ (\cos \theta + i \sin \theta)(\cos \theta - i \sin \theta) = \cos^2 \theta + \sin^2 \theta = 1.$$

Now suppose that we wish to multiply the complex numbers $z = r(\cos \theta + i \sin \theta)$ and $z' = r'(\cos \theta' + i \sin \theta')$. Then

$$zz' = rr'[(\cos \theta \cos \theta' - \sin \theta \sin \theta') + i(\cos \theta \sin \theta' + \sin \theta \cos \theta')].$$

Since the sine and cosine functions satisfy the fundamental addition relations

$$\cos(\theta + \theta') = \cos \theta \cos \theta' - \sin \theta \sin \theta'$$
$$\sin(\theta + \theta') = \cos \theta \sin \theta' + \sin \theta \cos \theta',$$

we infer that

$$zz' = rr'[\cos(\theta + \theta') + i \sin(\theta + \theta')]. \tag{6}$$

But this is the trigonometrical form of the complex number with modulus rr' and argument $\theta + \theta'$. Accordingly, *to multiply complex numbers one multiplies their moduli and adds their angles*[5]. Multiplication by a complex number of modulus 1 and argument θ thus corresponds precisely to rotation through the angle θ. In particular, *multiplication by* $i = \cos 90° + i \sin 90°$ *corresponds to rotation through a right angle.*

Taking $z = z'$ in (6), we get

$$z^2 = r^2(\cos 2\theta + i \sin 2\theta),$$

and, multiplying this result again by z,

$$z^3 = r^3(\cos 3\theta + i \sin 3\theta).$$

Continuing indefinitely in this way, we obtain, for arbitrary n,

---

[5] So the product of complex numbers is a "complex" of multiplication and addition.

$$z^n = r^n(\cos n\theta + i \sin n\theta).$$

Putting $r = 1$, we obtain the memorable formula of *A. De Moivre* (1667–1754)

$$(\cos \theta + i \sin \theta)^n = \cos n\theta + i \sin n\theta.$$

Using De Moivre's formula we can determine the roots of the equation $x^n - 1 = 0$—the nth *roots of unity*—in the field of complex numbers. Taking $\theta = m.360°/n$ for $m = 1, 2, ..., n$, we see that $n\theta$ is a multiple of $360°$, so that $\cos n\theta = 1$, $\sin n\theta = 0$. The formula then gives

$$(\cos \theta + i \sin \theta)^n = \cos n\theta + i \sin n\theta = 1 + 0i = 1.$$

Accordingly the roots of the equation $x^n - 1 = 0$ are

$$x = \cos(m.360°/n) + i \sin(m.360°/n)$$

for $m = 1, 2, 3, ..., n$. Writing α for the root $\cos(360°/n) + i \sin(360°/n)$, we see that the roots may be represented as $1, \alpha^2, ..., \alpha^{n-1}$ (in general, this is true for any root $\alpha \neq 1$.) Geometrically the values of x are represented by the vertices of the regular n–sided polygon inscribed in the unit circle, and so the equation $x^n - 1 = 0$ is known as the *cyclotomic*—"circle cutting"—*equation*.

THE THEORY OF NUMBERS

Number theory, hailed by Gauss as the Queen of Mathematics, abounds in problems that are easy to state, but extremely difficult to solve, and which have been the source of some of the deepest mathematical investigations. We outline a few of these problems, all of which have intrigued mathematicians for centuries.

Perfect Numbers

We recall that a number[6] is *perfect* if it is the sum of its proper divisors. The first four perfect numbers, viz., 6, 28, 496, 8128 were known to the Greeks, and the next, 33550336, appears in a medieval manuscript. In Book IX of Euclid's *Elements* it is proved (expressed in modern symbolism) that if $2^k - 1$ is prime, then $2^{k-1}(2^k - 1)$ is perfect. In the eighteenth century Euler proved what amounts to a converse for *even*

---

[6]Throughout our discussion of number theory, the term "number" or "integer" will always mean "natural number".

perfect numbers, namely, that any such number is of the form $2^{p-1}(2^p - 1)$, where both p and $2^p - 1$ are prime. Prime numbers of the form $2^p - 1$ are called *Mersenne primes* after *Marin Mersenne* (1588–1648). The complete list of currently known prime numbers p such that $2^p - 1$ is (a Mersenne) prime is: 2, 3, 5, 7, 3, 17, 19, 31, 61, 89, 107, 127, 521, 607, 1279, 2203, 2281, 3217, 4253, 4423, 9689, 9941, 11213, 19937, 21701, 23209, 44497, 86243, 32049, 216091, 756839, 859433, 1257787, 1398269, 2976221, 3021377, 6972593; to each of these there corresponds a perfect number. Thus the largest perfect number known (as of 1999) is

$$2^{6972592}(2^{6972593} - 1).$$

One may naturally ask: do the perfect numbers go on forever? Or is there a largest one? The answer to this question is still unknown. Another question which remains unanswered is: do *odd* perfect numbers exist? One of the few facts that has been established about these elusive numbers is that none smaller than 10^{200} can exist.

Prime Numbers

Prime numbers play the same role with respect to multiplication as does the number 1 with respect to addition: just as every number is uniquely expressible as a sum of 1s, so every number is uniquely expressible as a product of primes. (Of course, the first fact is trivial, but the second—the *fundamental theorem of arithmetic*—is not.) We have already pointed out that the Greeks knew that the sequence of primes

$$2, 3, 5, 7, 11, 13, 17, 19, 23, \ldots$$

is unending. As we recall, this is proved by showing that, for any number n, there is a prime between n and $n! + 1$ (where $n! = 1 \times 2 \times 3 \times \ldots \times n$). In 1850 the result was greatly improved by the Russian mathematician *P. Chebychev* (1821–1894) who showed that, for any number $n \geq 2$, there is always a prime between n and $2n$. This is the case despite the fact that, as we proceed through the number sequence, the primes become very sparsely distributed indeed. This becomes apparent when it is observed that, for any number n, we can find a sequence of n *consecutive* numbers none of which is prime. (Consider, for example, the sequence $(n+2)!+2, (n+2)!+3,\ldots, (n+2)!+(n+1)$.)

  Two famous results concerning prime numbers are *Fermat's* and *Wilson's Theorems*. These are most conveniently stated in terms of the idea of *congruence* of numbers. If m is a positive integer, and a,b integers (positive, negative, or zero), we say that a is *congruent* to b modulo m and write

$$a \equiv b \pmod{m}$$

if m divides $a - b$. It is readily shown that $a \equiv b \pmod{m}$ precisely when a and b leave the same remainder on division by m. *Fermat's Theorem*, stated in 1640 by *Pierre de Fermat* (1601–1665) is the assertion that, if p is prime, then for any number a,

$$a^p \equiv a \pmod{p}.$$

Wilson's Theorem (actually proved by Lagrange) is the assertion that, for any number n, n is prime if and only if $(n - 1)! \equiv -1 \pmod{n}$. [7]

Many attempts have been made to find simple arithmetical formulas which yield *only* primes. For example, in the seventeenth century Fermat advanced the famous conjecture that all numbers of the form

$$F(n) = 2^{G(n)} + 1,$$

where $G(n) = 2^n$, are prime. Indeed, for $n = 1, 2, 3, 4$ we have $F(1) = 5$, $F(2) = 17$, $F(3) = 257$, $F(4) = 65537$, all of which are prime. However, in 1732 Euler discovered the factorization $F(5) = 641 \times 6700417$, so that $F(5)$ is not a prime. And it is now known that all $F(n)$ with $5 \le n \le 19$ are composite (i.e., are not prime). So it is possible, although not so far established, that $F(n)$ is composite for all $n \ge 5$, and Fermat (almost) totally wrong.

Euler discovered the remarkable polynomial $n^2 - n + 41$, which yields primes for $n = 1, 2,..., 40$. The polynomial $n^2 - 79n + 1601$ yields primes for all values of n below 80. In the nineteen seventies explicit polynomials (in several variables) were constructed whose values comprise *all* the prime numbers. Thus, even though these polynomials are too complex to be of any practical use, in a formal sense the dream of number theorists of producing an algebraic formula yielding all the primes has finally been realized.

A decisive step was taken in the investigation of prime numbers when attention shifted from the problem of finding exact mathematical formulas yielding all the primes to the question of how the primes are, on the average, dispersed through the integers. While the primes are individually distributed with extreme irregularity, a remarkable regularity emerges when one considers the *likelihood* of a given number being prime. Writing $\pi(n)$ for the number of primes among the integers $1, 2, ..., n$, then the likelihood or *probability* [8] that a number selected at random from the first n integers will be prime may be identified with the quotient $\pi(n)/n$. The *prime number theorem* which expresses the regular behaviour of this quotient is regarded as one of the greatest discoveries in mathematics. In order to state it we need to define the concept of *natural logarithm* of a number. To do this we take two perpendicular axes in a plane and consider the curve comprising all points in the plane the product of whose distances x and y from these axes is equal to 1. In terms of the coordinates (x, y) of the point, this curve—a rectangular hyperbola—has equation $xy = 1$ and looks like this:

---

[7] Proving this theorem in one direction is easy, for if n has a factor d with $1 < d < n$, then d cannot divide $(n - 1)! + 1$, and so neither can n.

[8] If a given event can occur in m ways and fail to occur in n ways, and if each of the $m + n$ ways are equally likely, the *probability* of the event occurring is defined to be $m/(m + n)$.

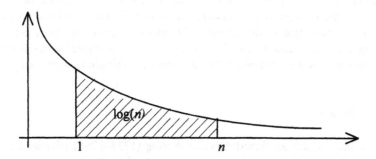

We now define $\log(n)$, the natural logarithm of n, to be the area in this figure bounded by the hyperbola, the x-axis, and the vertical lines $x = 1$ and $x = n$. In the late eighteenth century, through studying tables of prime numbers, Gauss observed that the quotient $\pi(n)/n$ is roughly equal to $1/\log(n)$, and that the approximation appears to improve with increasing n. Such empirical evidence led him to conjecture that $\pi(n)/n$ is "asymptotically equal" to $1/\log(n)$. By this we mean that the ratio of these two quantities, that is,

$$\frac{\pi(n)/n}{1/\log(n)} = \frac{\pi(n)}{n/\log(n)}$$

can be made to come as close to 1 as we please by making n sufficiently large. Although easily understood, proving Gauss' conjecture turned out to be extremely difficult, and indeed a rigorous proof was not forthcoming until 1896, nearly a century later.

Thus the problem of the average distribution of prime numbers has achieved a satisfactory solution. There remain, however, many other conjectures concerning prime numbers whose truth is suggested by empirical evidence but which have so far resisted final proof (or refutation).

One of these is the celebrated *Goldbach conjecture*. In a letter to Euler written in 1742 the amateur mathematician C. Goldbach (1690–1764) observed that every even number (except 2) that he had tested could be expressed as the sum of two primes: for example, $4 = 2 + 2$, $12 = 5 + 7$, $48 = 29 + 19$, etc. Goldbach asked Euler whether he could prove this to be true for all even numbers, or if he could provide an example to refute it. Euler never came up with an answer, and the problem—Goldbach's conjecture—still awaits solution. The empirical evidence for the truth of this conjecture is very strong, but the fact that it involves the *addition* of primes, while these themselves are defined in terms of *multiplication*, makes the construction of a rigorous proof no easy matter. In fact, it is rarely simple to establish connections between the multiplicative and additive properties of the integers. Nevertheless, some progress with Goldbach's problem has been made: we now know, for example, that every *sufficiently large*[9] even number may be expressed as the sum of no more than four primes, and also as the sum of a prime and a number with no more than two prime factors.

A final "prime riddle" on which little or no progress has been made is the *twin prime conjecture*, that is, the assertion that there are arbitrarily large prime numbers *p* for which *p* + 2 is also prime. This problem differs in logical status from Goldbach's conjecture in that it cannot be refuted by supplying a single counterexample.

Sums of Powers

In 1770 the English algebraist *Edward Waring* (1734–1798) advanced the claim that every number was the sum of four squares, nine cubes, 19 fourth powers, etc. This was pure conjecture on his part, but not long afterwards Lagrange succeeded in establishing the correctness of Waring's claim for squares (a claim which had actually been made much earlier by Fermat). It was not until 1909, however, that Waring's assertion was shown (by Hilbert) to be true in the general sense that, for any number *k*, *there exists* a number *s* such that every integer can be expressed as a sum of no more than s k^{th} powers. The problem of determining the least such number $w(k)$ for arbitrary *k* is known as *Waring's problem*. It is now known how to calculate $w(k)$ for all *k*, with the exception, curiously, of $k = 4$, although it is known that $w(4) \leq 19$ (so that Waring was right in this case). For all *k* such that $1 \leq k \leq 200000$, except $k = 4$, it turns out that

$$w(k) = 2^k - 2 + \text{the largest integer} < (3/2)^k.$$

Now although $w(3) = 9$, very few integers actually require as many as 9 cubes to represent them: in fact, only 23 and 239 require so many. The largest integer requiring eight cubes is 454, and inspection reveals that the proportion of integers requiring seven cubes decreases as we proceed through them. Numbers like 23, 239, 454 thus seem to be no more than irritating anomalies. In view of this it is more interesting to consider, instead of $w(k)$, the number $W(k)$ defined to be the least value of *s* for which every *sufficiently large* integer can be written as a sum of no more than s k^{th} powers. It is known that $W(2) = 4$, and, strangely, the only other value of *k* for which $W(k)$ is known with certainty is $k = 4$, the one value of *k* for which $w(k)$ is unknown. In fact $W(4) = 16$; all that is known about $W(k)$ for other values of *k* is that they cannot exceed a certain size: for example, $W(3) \leq 7$ and $W(5) \leq 23$. The most that is known in general about $W(k)$ is that

$$k + 1 \leq W(k) \leq k(3\log(k) + 11),$$

the right hand inequality being extremely difficult to prove. Waring's problem is still an active issue in number theory.

---

[9] By a "sufficiently large" number of a given type we mean a number of that type exceeding some fixed number specified in advance.

Fermat's Last Theorem

It was Fermat's habit to record his observations in the margins of his copies of mathematical works. In commenting on a problem in Diophantus's *Arithmetic*, asking for the solution in rational numbers of the equation $x^2 + y^2 = a^2$, Fermat remarked that he had found a "truly marvellous demonstration" of the assertion that, by contrast, there are no integral solutions to any of the equations $x^n + y^n = z^n$ for $n \geq 3$, adding that, unfortunately, the margin was "too narrow to contain" the demonstration. This assertion is often called *Fermat's Last Theorem*, but it would be more apt to term it "Fermat's conjecture", since it is not known whether he actually had a correct proof of it. It seems unlikely that he did since the best efforts of his successors to prove it failed until very recently. Fermat himself gave an explicit proof for the case $n = 4$, and Euler proved the case $n = 3$ between 1753 and 1770 (a lacuna in the proof later being filled by Legendre). Around 1825 proofs for $n = 5$ were independently formulated by Legendre and Dirichlet, and in 1839 Lamé proved the theorem for $n = 7$. In the nineteenth century significant advances in the study of the problem were made by the German mathematician E. Kummer, which led to the development of the theory of *algebraic numbers* (*q.v.* Chapter 6). Before the First World War, a substantial prize (the value of which was subsequently wiped out by inflation) was offered in Germany for a complete proof, and many amateurs contributed attempted solutions. Legend has it that the number theorist *Edmund Landau* (1877–1938) had postcards printed which read: "Dear Sir or Madam: Your attempted proof of Fermat's Theorem has been received and is hereby returned. The first mistake occurs on page ___, line ___." These he would give to his students to fill in the missing numbers. In 1994, the efforts of mathematicians of the last three centuries to prove Fermat's Theorem culminated in Andrew Wiles' successful complete proof. This is of great depth and complexity, drawing on results and techniques from several areas of mathematics ouside number theory which had not been developed in Fermat's day. Nevertheless, we cannot be entirely certain that Fermat—a great mathematician—did not possess a proof himself, one which his successors have so far failed to see.

The number π

The real number π is defined[10] to be the length of the circumference of a circle of unit diameter, and is thus the ratio of the length of the circumference of any circle to its diameter. It may also be defined as the area of a circle of unit radius, since, as was known to the Greeks, the area of a circle is equal to one half of the product of its radius by its circumference. It is of interest to see how it is believed this result was first discovered. If a circle of radius r and circumference c is divided into a large number of

---

[10]The general use of the symbol π in its familiar mathematical sense was first adopted by Euler in 1737.

segments, as in the figure below, then, if there are sufficiently many of them, each may be taken as an actual triangle. In other words, we may take the circle to be a regular polygon with a large number of sides. This polygon can now be opened out to give a

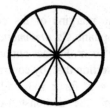 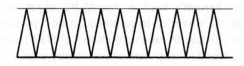

serrated figure in which the height of each serration is r and the length of the base is c. But clearly the area of this figure, and hence also of the original circle, is one half of the area of the rectangle with the same base and height, i.e. $\frac{1}{2}rc$.

The problem of computing π has a very long history. In ancient Egypt its value was often taken as 3, but the better value $(4/3)^4 = 3.1604...$ appears in the Rhind Papyrus. The first truly scientific attempt to compute π seems, however, to have been made by Archimedes around 240 B.C. In his work *The Measurement of a Circle*, (mentioned in the previous chapter) by evaluating the perimeters of regular inscribed and circumscribed polygons he shows that π falls between 223/71 and 22/7, so that, to two decimal places, $\pi = 3.14$. The first value of π better than that of Archimedes was obtained by *Ptolemy* ($c.85$–165) around 150 A.D. in his great astronomical work popularly known through its Arabic name *The Almagest*. Ptolemy obtained the value 377/120 = 3.1416. Around 480 A.D. the Chinese mathematician *Tsu Ch'ung-chih* (430–501) obtained the unusual rational approximation 355/113 = 3.1415929, correct to six decimal places, and in 530 the Hindu mathematician *Aryabhata* ($c.475$–550) gave the fraction 62832/20000 = 3.1416 as an approximate value for π. In 1150 the Hindu mathematician *Bhaskara* (1114–1185) produced several approximations for π: 3927/1250 as an accurate value, and $\sqrt{10}$ for ordinary calculations, among others.

In 1579 *François Viète* (1540-1603) determined π correct to nine decimal places by applying Archimedes' method to polygons having $6.2^{16} = 393296$ sides. Also due to him is the curious "infinite product" in which, by taking the product of sufficiently many terms on the right, we may approximate to π as closely as we please:

$$2 \cdot \frac{2}{\sqrt{2}} \cdot \frac{2}{\sqrt{(2+\sqrt{2})}} \cdot \frac{2}{\sqrt{(2+\sqrt{(2+\sqrt{2})})}} \cdots$$

In 1650 *John Wallis* (1616-1703) also obtained π in the form of an infinite product

$$\frac{1}{2}\pi = \frac{2.2.4.4.6.6.8...}{1.3.3.5.5.7.7...},$$

which was converted by *Lord Brouncker* (c.1620-1684) into the "continued fraction"

$$4/\pi = 1 + \cfrac{1^2}{2 + \cfrac{3^2}{2 + \cfrac{5^2}{2 + ...}}}$$

In 1671 *James Gregory* (1638–1675) obtained π as an "infinite series" (see Chapter 9)

$$\tfrac{1}{4}\pi = 1 - 1/3 + 1/5 - 1/7 + ...,$$

in which, by adding and subtracting sufficiently many terms on the right, we can approximate to $\tfrac{1}{4}\pi$ as closely as we please.

Other remarkable series involving π are those obtained in the eighteenth century by Euler:

$$\pi^2/6 = 1 + 1/2^2 + 1/3^2 + 1/4^2 + ...$$

and

$$\pi^2/8 = 1 + 1/3^2 + 1/5^2 + 1/7^2 + ...$$

A major step in the elucidation of the nature of π was taken in 1767 by *Johann Heinrich Lambert* (1728–1777) when he established its *irrationality*. In 1794 *Adrien-Marie Legendre* (1752–1833) took this further by showing that π^2 is irrational, so that π cannot be the square root of a rational number.

An intriguing connection between π and the concept of *probability* was noted in 1777 by the *Comte de Buffon* (1707–1788) through the devising of his famous *needle experiment*. Buffon showed that, if a needle is dropped at random on a uniform parallel ruled surface on which the distance between successive lines coincides with the length of the needle, then the probability of the needle falling across one of the lines is $2/\pi$. The incongruous presence of π here results from the fact that the outcomes of the experiment depend not only on the fallen needle's position, but also on the *angle* it makes with the lines on the surface, and angles are measured by lengths of circular arcs, i.e. in terms of π. By noting the number of successful outcomes of a large number of trials of the experiment, one obtains a value for the probability and hence also an approximate value for π . In 1901 the experiment performed with 3400 tosses of the needle led to an amazingly accurate estimate of π correct to six decimal places!

In 1882 *C. L. F. Lindemann* (1852–1939) proved the decisive result that π is *transcendental*, that is, not a root of any polynomial with integer coefficients. This showed conclusively that the ancient problem of squaring the circle cannot be solved with Euclidean tools.

WHAT ARE NUMBERS?

While there is universal agreement on the rules for calculating with the natural numbers, there has been a surprising lack of unanimity concerning what they actually *are*. To Aristotle, for example, a number was an *arithmos*, a plurality of definite things, a collection of indivisible "units." The Greeks did not regard 1 (let alone 0) as a "number" since it is a unit rather than a plurality. Neither could they conceive of fractions as numbers since the unit is indivisible: numbers for them were discrete entities to be distinguished absolutely from geometric magnitudes which are continuous and can be divided indefinitely. Although Diophantus in the third century B.C. had suggested that the unit be treated as divisible to facilitate the solution of certain problems, it was not until the sixteenth century that the Greek concept of number as an assemblage of discrete units began seriously to give way to the idea of number as a *symbol indicating quantity in general*, including continuous quantity. Thus *Simon Stevin* (1548–1620) avers that number is not to be identified with discontinuous quantity, and that to a continuous magnitude there corresponds a continuous number. Stevin regards not only 0 and 1, but fractions and even irrational numbers such as $\sqrt{2}$ as "numbers." His successors were to take up this idea with gusto.

Although this enlargement of the concept of number proved a valuable stimulus for the development of mathematics, it had at the same time the paradoxical effect of rendering obscure the nature of the original "numbers"—the *natural numbers*—that had given birth to the new concepts. For while fractions, negative numbers and the like are essentially symbols signalizing the effect of *operations*[11], the natural numbers seem to have a more immediate, concrete, even eidetic character—as the Greeks acknowledged in their identification of numbers with *arithmoi*.

In the last quarter of the nineteenth century these issues excited a new interest among mathematicians, leading to the publication of several works in which attempts were made to clarify the nature of number, and to put the whole subject of arithmetic on a logical basis. The first of these to appear was *Gottlob Frege's* (1848 – 1925) *The Foundations of Arithmetic*, which was published in 1884. This book has been called the first philosophically sound discussion of the concept of number in Western civilization, an assessment with which it would be hard to disagree. In it Frege subjects the views on the nature of number of his predecessors and contemporaries to merciless analysis, finally rejecting them all, and proposes in their place his own compellingly subtle theory. It is worth quoting his summary of the difficulties standing in the way of arriving at a satisfactory account of number.

> Number is not abstracted from things in the way that colour, weight and hardness are, nor
> is it a property of things in the sense that they are. But when we make an assertion of
> number, what is that of which we assert something? This question remained unanswered.
> Number is not anything physical, but nor is it anything subjective (an idea).

---

[11] It was the fact that imaginary and complex numbers could not (at first) be conceived of as operations that prevented them from being regarded as "numbers" —even in this extended sense—until the end of the eighteenth century.

Number does not result from the annexing of thing to thing. It makes no difference even if we assign a fresh name to each act of annexation.

The terms "multitude", "set" and "plurality" are unsuitable, owing to their vagueness, for use in defining number.

In considering the terms one and unit, we left unanswered the question: How are we to curb the arbitrariness of our ways of regarding things, which threatens to obliterate every distinction between one and many?

Being isolated, being undivided, being incapable of dissection—none of these can serve as a criterion for what we express by the word "one".

If we call the things to be counted units, then the assertion that units are identical is, if made without qualification, false. That they are identical in this respect or that is true enough but of no interest. It is actually necessary that the things to be counted should be different if number is to get beyond 1.

We [are] thus forced, it seem[s], to ascribe to units two contradictory qualities, namely identity and distinguishability.

A distinction must be made between one and unit. The word "one", as the proper name of an object of mathematical study, does not admit of a plural. Consequently, it is nonsense to make numbers result from the putting together of ones. The plus symbol in 1 + 1 = 2 cannot mean such a putting together.

The account of number Frege puts forward in *The Foundations of Arithmetic* has several key features. First, by "number" he means *cardinal number*, that is numbers such as *one, two, three* which answer to the question "how many?", as opposed to *ordinal number* such as *first, second, third* which answer to the question "what position in a series?". Secondly, and relatedly, numbers are to be treated as definite *objects*, rather than predicates or properties. Finally, and crucially, numbers are to be conceived as attaching not directly to things, but rather to *concepts*. Thus, for example, in saying that there are five fingers on my right hand I am assigning the number five to the concept "finger on my right hand", rather than to the actual assemblage of fingers itself. Frege thus maintains that numbers only become assigned to things in an indirect way: first the *concept* of the thing is abstracted from the given things, and then the *number* is assigned to the concept. Frege suggests that the concept itself is the "unit" in respect of the number assigned to it. Thus, for example, in saying that there are five fingers on my right hand the relevant unit is the concept "finger", but in asserting that there are fourteen joints on the same hand the unit is the concept "joint". This, he claims, makes it easy to reconcile the identity of units with their distinguishability, for here the word "unit" is being used in a double sense. First, it is used in the sense of "concept", and since one single concept, for example "finger", attaches to the collection to which the number is being assigned, the "units" in this sense are identical. On the other hand, when we assert that units are distinguishable, we mean "unit" in the sense of "thing numbered", and these are, for the purpose of numeration, always taken as distinct.

Thus Frege has explicated how numbers come to be *assigned*, but he has not yet determined what they actually *are*. To do this he employs the familiar fact that two collections of things have the same number precisely when the members of one collection can be paired off with those of the other in such a way that both are exhausted: in this event we say that the two collections are *equinumerous*. This term may be extended to concepts by saying that two concepts are equinumerous if the collections of objects to which the concepts respectively attach are equinumerous. Thus, for example, the concepts "side of a triangle" and "vertices of a triangle" are equinumerous in this sense. Frege then defines the *number* assigned to a given concept to be the collection of all concepts equinumerous with the given one[12]. Thus for Frege a number corresponds to a *collection of concepts*. The specific natural numbers 0, 1, 2,... can then be defined as follows:

> 0 is the number assigned to the concept "not identical with itself"[13];
> 1 is the number assigned to the concept "identical with 0";
> 2 is the number assigned to the concept "identical with 0 or identical with 1"
> ....
> *n* is the number assigned to the concept "identical with 0 or identical with 1 or identical with 2 or ... or identical with $n - 1$".

Frege admits that it may seem somewhat strange to define a number as a collection of concepts, but he goes on to show that from his definition all the usual facts about numbers (including the Principle of Mathematical Induction) can be derived. It was highly unfortunate that one of the logical assumptions on which Frege's account of number rested turned out to be inconsistent; however, later investigations have shown that his framework can be salvaged: see Chapter 12.

In 1888 a second major work on the foundations of arithmetic was published—*The Nature and Meaning of Numbers* by the German mathematician *Richard Dedekind* (1831–1916). Unlike Frege, Dedekind is not concerned to give a completely explicit definition of natural number itself; instead, he specifies a structure (based on the primitive notion of *set* or *class*) which possesses the essential properties of the *whole sequence* of natural numbers, and then obtains the natural numbers themselves by an act of mental abstraction.

To grasp the underlying motivation of Dedekind's approach, consider the set **N** of natural numbers $\{1, 2, 3,...\}$, where, following Dedekind, we take the sequence as beginning with 1. With each number *n* we associate its successor $n' = n + 1$. Dedekind generalizes this in the following way. Instead of **N** he takes an arbitrary set *S* (which he calls a "system") and instead of the successor operation an arbitrary one-one function φ from *S* into itself (for definitions of these terms, see Chapter 4). Now, for an arbitrary element *s* of *S*, write s' for $\varphi(s)$ and if *A* is any subset of *S*, write A' for the set consisting of all elements of the form s' for *s* in *A*. Clearly **N**′ is included in **N**. Dedekind accordingly defines a *chain* to be a subset *K* of *S* such that K' is included in *K*. Since, intuitively, **N** is the *smallest* chain which contains the number 1, he defines by analogy the *chain* of a subset *A* of *S*, written A_0, to be the smallest chain which includes *A*, that is, the common part of all chains which include *A*. Dedekind next introduces the concept of *simply infinite system*: by this he means a set *S* together with an operation φ of *S* into itself and an element 1 of *S* such that: (i) $S = 1_0$, (ii) the element 1 is not contained in S', (iii) the operation φ is one-one. Finally he makes the

---

[12] Strictly speaking, Frege defines this number to be what is known as the *extension* of the concept "equinumerous to the given concept": see Chapter 12 for further details.

[13] It will be seen that the concept "not identical with itself" attaches to no object, so that the corresponding number is indeed zero.

Definition. If in the consideration of a simply infinite system N set in order by a transformation φ we entirely neglect the special character of the elements; simply retaining their distinguishability and taking into account only the relation to one another in which they are placed by the order-setting transformation φ, then are these elements called *natural numbers* or *ordinal numbers* or simply *numbers*, and the base element 1 is called the *base-number* of the *number-series* N. With reference to this freeing the elements from every other content (abstraction) we are justified in calling numbers a free creation of the human mind. The relations or laws which are derived entirely from the conditions [(i), (ii), (iii) above] and therefore are always the same in all ordered simply infinite systems, whatever names may happen to be given to the individual elements ... form the first object of the *science of numbers* or arithmetic.

The Principle of Mathematical Induction is now an immediate consequence of this definition of natural numbers. For if P is any property defined on the members of a simply infinite system N such that 1 has P, and, whenever a has P, so does a', then clearly the set K of elements having P is a chain containing 1. Since $N = 1_0$ is the least chain containing 1, it must be included in K, so that every member of N has P.

As we have seen, Dedekind obtains the natural numbers by "neglecting the special character" of the elements of a simply infinite system. But how can one be certain that simply infinite systems exist in the first place? Dedekind shows that a simply infinite system can always be obtained from an *infinite* system (or set) which he famously defines as follows:

A system S is said to be *infinite* when it is similar to a proper part of itself...; in the contrary case S is said to be a *finite* system.

Here two sets are said to be *similar* when they can be put in one-one correspondence. Dedekind offers the following somewhat curious proof of the existence of infinite sets:

Theorem. There exist infinite systems.
Proof. My own realm of thoughts, i.e., the totality S of all things, which can be objects of my thought, is infinite. For if s signifies an element of S, then is the thought s', that s can be the object of my thought, itself an element of S.

The correspondence $s \rightarrow s'$ is a one-one correspondence between S and a proper part of itself, and so S is infinite.

Most mathematicians were not persuaded by this "proof", and it later came to be seen that the existence of infinite sets is a matter of postulation rather than proof. (In Frege's system, however, the existence of an infinite set can actually be proved.)

The period 1894–1908 saw the publication of the five volumes of the Italian mathematician *Giuseppe Peano*'s (1858–1932) *Formulaire de Mathématique*, a series of works devoted to presenting the fundamental concepts of mathematics in a rigorous symbolic form. Volume III, published in 1901, contains a version of Peano's treatment of the foundations of arithmetic, which while itself owing much to Dedekind, has proved definitive. Peano bases his formulation on three primitive ideas: (i) N_0, the class of natural numbers, (ii) 0, the particular number zero, (iii) $a+$, the successor of the number a. He lays down the the following five postulates—the well-known *Peano postulates* for the natural numbers:

1. 0 is a number.
2. The successor of any number is a number.
3. If a class S is such that (a) it contains 0 and (b) if it contains any number a it also contains the successor $a+$ of that number, then S includes the whole of N_0.
4. No two numbers have the same successor.
5. 0 is not the successor of any number.

The third of these postulates is the Principle of Mathematical Induction.

Peano's approach differs both from Frege's and Dedekind's in that, rather than attempting to define the natural numbers in terms of something more primitive, he simply takes natural number as an undefined notion to be characterized *axiomatically*.

In 1923 the Hungarian mathematician *John von Neumann* (1903–1957) formulated a definition of (ordinal) number within set theory which has since become standard. His idea was similar to, but simpler than Frege's: whereas for Frege each number n is the number assigned to the *concept* of being among the set of numbers $\{0, 1,..., n-1\}$, von Neumann simply defines each n to be *identical* with the set $\{0, 1, ..., n-1\}$. In other words von Neumann simply identifies each number with the set of its predecessors. That being the case, the number 0, lacking predecessors altogether, must be identical with the empty set $\varnothing$, 1 must be the set $\{\varnothing\}$, 2 the set $\{\varnothing,\{\varnothing\}\}$, 3 the set $\{\varnothing,\{\varnothing\},\{\varnothing,\{\varnothing\}\}\}$, etc.

*

If we survey the evolution of the number concept, we see that at each crucial stage a new kind of symbolic "number" was created so as to enable a certain sort of equation to be solved. We conclude this chapter by displaying the resulting correspondence between "numbers" and equations as a scheme in which the entries in the left hand column indicate the type of number, and those in the right the simplest equation which has that type of number as a solution.

Rational numbers	$2x = 1$
Irrational numbers	$x^2 = 2$
Zero	$x + 1 = 1$
Negative numbers	$x + 2 = 1$
Complex numbers	$x^2 + 1 = 0$

In Chapter 11 we shall discuss infinite numbers, which may be brought into the scheme as

Infinite numbers	$x + 1 = x.$

CHAPTER 4

THE EVOLUTION OF ALGEBRA, I

TRADITIONALLY, ALGEBRA WAS THE BRANCH of mathematics concerned with operations on—and equations involving—numbers. Taken in this sense, algebra is of great antiquity, since solutions to quadratic, cubic, and simultaneous equations are to be found inscribed on Babylonian stone tablets dating from 2000–1600 B.C.

Greek Algebra

The *Greek* mathematicians were primarily geometers, and it was accordingly natural that they should solve arithmetic and algebraic problems by means of geometric constructions, for example, by producing line segments whose lengths correspond to roots of equations. In this manner Euclid, for instance, solved problems equivalent to solving the systems of simultaneous equations $xy = k^2$, $x^2 - y^2 = a$ and $xy = k^2$, $x \pm y = a$.

The first substantial work in which arithmetic is treated in a manner wholly independent of geometry is the *Introductio Arithmetica* of *Nicomachus of Gerasa* (*c.* 100 A.D.). In this work, which is chiefly devoted to the arithmetic of the early Pythagoreans, numbers are used to denote *quantities of objects—arithmoi—*and not line segments as in Euclid. The work was extremely popular and actually used as a schoolbook down to the Renaissance.

Like all its predecessors, the *Introductio* is a *rhetorical* work in that no use of symbols is made, the exposition proceeding entirely verbally. The first work of algebra to employ symbols in any significant way—the first *syncopated* algebra (from Greek *syncope*, "cut short")—was the *Arithmetica* of *Diophantus*, who flourished *c.*250 A.D. In this work Diophantus employs a character (probably the Greek letter ς, corresponding to our *x*) to represent the unknown quantity in an equation, and also uses initial letters to stand for equality and the operations of squaring and cubing.

The greater part of the *Arithmetica* is devoted to the solution of problems leading to linear and quadratic *indeterminate equations*, that is, equations without unique solutions. A few problems lead to equations of the third and fourth degrees. The general type of problem considered by Diophantus is to find two, three, or four numbers such that different expressions involving them in the first and second, and occasionally in the third degree, are squares, cubes, partly squares and partly cubes, etc. For example: *To find three numbers such the product of any pair added to the sum of the pair yields a square; To find four numbers such that, if we take the square of the sum ± any one of them singly, all the resulting numbers are squares; To find two numbers such that their product ± their sum gives a cube.*

Diophantus recognized only positive rational solutions to these equations, and for this reason algebraic problems in which only rational (or integral) solutions are considered are today termed *Diophantine*.

Chinese Algebra

Algebra played a major role in early *Chinese* mathematics. While Chinese algebra was chiefly rhetorical, employing symbols only rarely (and then only in its later period), in contrast with Greek practice solutions to geometric problems were often cast in algebraic form. An arresting example of this occurs in the treatment of Pythagoras's theorem to be found in the oldest of the Chinese mathematical classics, the *Chou Pei Suan Ching*—The Arithmetical Classic of the Gnomon and the Circular Paths of Heaven"—, which was probably composed during the Han period (206 B.C. – 222 A.D.). Here we see the square on the hypotenuse folded backwards[1] onto the

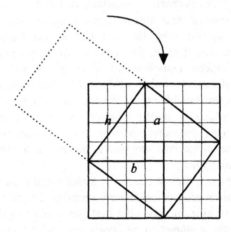

original triangle, manifestly containing three further identical triangles together with a square of side equal to the difference between the triangle's base and altitude. An algebraic formulation is given rhetorically in the text, which, in modern terms may be written

$$h^2 = 4ab/2 + (a - b)^2 = a^2 + b^2,$$

---

[1] It is possible that this demonstration was suggested by the Chinese art of paper-folding, since paper is believed to have been invented in China sometime before the first century B.C.

where h is the hypotenuse, a the altitude, and b the base. This demonstration is quite different from that of the Greeks.

The Chinese were adept at solving simultaneous linear equations, the coefficients of which they represented by means of rods on a counting-board, an arrangement precisely similar to that of numbers in a *matrix* (see below). The *Chiu Chang Suan Shu*—"Nine Chapters in the Mathematical Art", $c.100$ B.C.—contains numerous problems requiring the solution of simultaneous linear equations of the type

$$ax + by = c \qquad a'x + b'y = c'.$$

Here the first equation was multiplied by a' and the second by a, yielding, after subtraction,

$$y = \frac{ca' - c'a}{ba' - b'a},$$

this result, as usual, being expressed in words.

In the *Sun Tzu Suan Ching*—"Master Sun's Mathematical Manual"—of $c.400$ A.D. we find the following problem:

> We have a number of things, but we do not know exactly how many. If we count them by threes we have two left over. If we count them by fives we have three left over. If we count them by sevens we have two left over. How many things are there?

This is a problem in *number congruences,* which, written in modern form, may be expressed as follows: determine a number N such that

$$N \equiv 2 \; (\mathrm{mod}\ 3), \quad \equiv 3 \; (\mathrm{mod}\ 5), \quad \equiv 2 \; (\mathrm{mod}\ 7).$$

Sun Tzu gives 23 as a value for N, which is the least possible answer. This marks the beginnings of the famous *Chinese Remainder Theorem* of elementary number theory. The theorem states that, if we are given a set $\{m_1, ..., m_k\}$ of numbers no pair of which has a common factor apart from 1—such numbers are said to be *relatively prime*—then the system of congruences

$$x \equiv a_i \; (\mathrm{mod}\ m_i), \quad i = 1, 2, ..., k,$$

has a unique solution modulo $m_1 m_2...m_k$.

By the fourteenth century the Chinese had developed the technique for generating approximate solutions to algebraic equations known in the West as *Horner's method* (and published in 1819). For example, in the *Ssu-Yuan Yü Chien*—"Precious Mirror of the Four Elements", which appeared in 1303—we find an approximate solution to the equation $x^3 - 574 = 0$. First, noting that $x = 8$ is an approximate solution, $y = x - 8$ is substituted to yield the equation $y^3 + 24y^2 + 192y - 62 = 0$, which has a solution between 0 and 1. Taking y^3 and y^2 to be approximately 1 now gives $y = 62/(1 + 24 + 192) = 2/7$ so that $x = 58/7$.

In the "Precious Mirror" we also find a diagram of what is known in the West as *Pascal's triangle*, which gives the coefficients of binomial expansions $(a + b)^n$:

$$1$$
$$1 \ 1$$
$$1 \ 2 \ 1$$
$$1 \ 3 \ 3 \ 1$$
$$1 \ 4 \ 6 \ 4 \ 1$$
$$1 \ 5 \ 10 \ 10 \ 5 \ 1$$

etc.

The author of the "Precious Mirror" refers to this diagram as "the old method for finding eighth and lower powers", which shows that the Chinese must have already formulated the binomial theorem by the start of the twelfth century at the latest.

Hindu Algebra

Significant contributions to algebra were also made by *Hindu* mathematicians during the period 200–1200 A.D. Hindu algebra was very close to being syncopated: to describe operations they used abbreviations of words together with a few special symbols, and for unknowns in equations they used the names of colours—black, blue, yellow, etc. The initial letter of each colour word was also used as a symbol. The Hindu mathematicians realized that quadratic equations had two roots, and they accepted the presence of both negative and irrational roots. They also took the important step of extending to irrational numbers the operations performed on integers, although this procedure lacked rigorous justification. For example Bhaskara shows how to sum irrationals as follows. Given the irrationals $\sqrt{3}$ and $\sqrt{12}$, we get

$$\sqrt{3} + \sqrt{12} = \sqrt{[(3 + 12) + 2\sqrt{(3.12)}]} = \sqrt{27} = 3\sqrt{3}. \qquad (*)$$

Here Bhaskara is treating the irrationals as if they were integers, a fact he explicitly acknowledges. For if in the identity

$$m + n = \sqrt{(m + n)^2} = \sqrt{(m^2 + n^2 + 2mn)},$$

clearly true for integers m, n, we substitute $\sqrt{a}$ for m and $\sqrt{b}$ for n, we obtain

$$\sqrt{a} + \sqrt{b} = \sqrt{[\sqrt{a} + \sqrt{b} + 2\sqrt{(ab)}]}$$

of which (*) above is a special case.

The Hindus advanced well beyond Diophantus in their treatment of *indeterminate equations*, most of which derived from astronomical problems, their solutions representing the appearance of certain constellations. Where Diophantus sought just a single rational solution to these equations, the Hindus developed methods sought just a single rational solution to these equations, the Hindus developed methods for obtaining *all* integral solutions. Aryabhata devised a method—later employed by his successor *Brahmagupta* (fl. 628) for obtaining the integral solutions to the linear Diophantine equation $ax \pm by = c$, where a, b and c are positive integers. Let us outline this method, using modern symbolism, in the case of $ax + by = c$.

First, we find the greatest common divisor (a, b) of a and b by means of the *Euclidean algorithm*. This involves performing the successive divisions:

$$\begin{aligned}
a &= bq_1 + r_1 & (0 < r_1 < b) \\
b &= rq_2 + r_2 & (0 < r_2 < r_1) \\
r_1 &= r_2q_3 + r_3 & (0 < r_3 < r_2) \\
r_2 &= r_3q_4 + r_4 & (0 < r_4 < r_3) \\
&\;\;\vdots \\
&\;\;\vdots
\end{aligned} \tag{1}$$

so long as none of the remainders $r_1 > r_2 > r_3 > \dots$ are 0. After at most b steps the remainder 0 must appear:

$$\begin{aligned}
r_{n-2} &= r_{n-1}q_n + r_n \\
r_{n-1} &= r_n q_{n+1} + 0
\end{aligned}$$

Now from the successive lines of (1) it follows that $(a, b) = (b, r_1) = (r_1, r_2) = \dots$ $= (r_{n-1}, r_n) = (r_n, 0) = r_n$. Therefore (a, b) is the last positive remainder in the sequence $r_1 > r_2 > r_3 > \dots$. It also follows from (1) that if $d = (a, b)$, then (positive or negative) integers k and ℓ can be found such that

$$d = ka + \ell b \tag{2}$$

For using the first equation in (1) we get

$$r_1 = a - q_1 b,$$

so that r_1 can be written in the form $k_1 a + \ell_1 b$. From the next equation, we obtain

$$r_2 = b - q_2 r_1 = b - q_2(k_1 a + \ell_1 b) = -q_2 k_1 a + (1 - q_2 \ell_1)b = k_2 a + \ell_2 b.$$

This procedure can clearly be iterated through the successive remainders r_3, r_4, ... until we arrive at a representation

$$r_n = ka + \ell b,$$

as required.

According to (2), our original equation $ax + by = c$ has the particular solution $x = k$, $y = \ell$ for the case $c = d$. In general, if $c = d.q$ is any multiple of d, then from (2) we get

$$a(kq) + b(\ell q) = dq,$$

so that our original equation has the particular solution $x = x^* = kq$, $y = y^* = \ell q$. Conversely, if our equation has a solution x, y for a given c, then any solution must be a multiple of $d = (a, b)$, for d divides a and b, and so must also divide c. Accordingly, we see that our equation has a solution if and only if c is a multiple of (a, b).

To determine the remaining solutions, we observe that if $x = x'$, $y = y'$ is any solution other than the one $x = x^*$, $y = y^*$ just found by the use of the Euclidean algorithm, then $x = x' - x^*$, $y = y' - y^*$ is easily seen to be a solution to the equation

$$ax + by = 0. \tag{3}$$

Now the most general solution to (3) is

$$x = rb/(a, b), \quad y = -ra/(a, b).$$

For, dividing (3) by $(a, b) = d$, we obtain

$$a'x + b'y = 0,$$

where $a' = a/d$ and $b' = b/d$ are relatively prime. Therefore

$$a'x = -b'y \tag{4}$$

and since b' is relatively prime to a', it must divide x, i.e., $x = rb'$ for some integer r. From (4) it now follows that $y = -ra'$, as claimed.

Accordingly the general solution to our original equation, assuming c to be a multiple of (a, b) is

$$x = x^* + rb/(a, b), \quad y = y^* - ra/(a, b).$$

This is essentially Brahmagupta's solution.

Brahmagupta also considered the Diophantine quadratic equation[2] $x^2 = 1 + py^2$. Later the medieval mathematician Bhaskara, in a remarkably impressive feat of calculation, furnished particular solutions to this equation for $p = 8, 11, 32, 61$ and 67.

Arabic Algebra

Algebra entered Arabic (or Moslem) culture through the medium of a treatise on equations—based in the main on the work of Brahmagupta, but also showing traces of Greek and Babylonian influence—written in 830 by the mathematician *Mohammed ibn*

---

[2] This equation was mistakenly named for the English mathematician *John Pell* (1611–1685), but was actually first considered by Archimedes.

Musa al-Khwarizmi. It is from the title of this treatise, *Al-jabr w'al muqâbala*, which in free translation means "Restoring and simplification", that the word "algebra" originates[3]. Here the word "al-jabr" meant restoring the balance of an equation by adding or subtracting on one side a term which had been removed from the other, as in transforming $x^2 + 4 = 6$ into $x^2 = 6 - 4 = 2$. The word "al-muqâbala" meant simplification of an equation in the sense of subtracting equal terms from both of its sides or combining several similar terms into a single term, for example $2x^2$ and $4x^2$ into $6x^2$.

The algebra of al-Khwarizmi holds a most important position in the history of mathematics, since the subsequent Arabic and medieval works on algebra were founded on it, and, moreover, it served as the conduit through which the Hindu-Arabic system of decimal notation was introduced into the West. However, in certain respects al-Khwarizmi's work is a backward step from that of Brahmagupta, and even from that of Diophantus. The problems treated are far more elementary than those discussed by the latter, and in particular there is little discussion of indeterminate equations. Also al-Khwarizmi's algebra is entirely rhetorical. Nevertheless, the straightforward character and clear argumentation of the work makes it very much a forerunner of today's school algebra texts.

In his algebra al-Khwarizmi solves linear and quadratic equations by "completing the square" and recognizes that the latter can have two roots[4]—but at the same time he rejects negative roots. He also solves a number of quadratic equations by means of geometric arguments. For instance, to solve the equation $x^2 + 10x = 39$ he draws a square *ab* to represent x^2, and on the four sides of this square he places

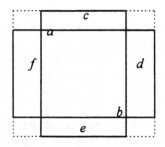

rectangles *c, d, e,* and *f,* each 2½ units wide. To complete the larger square, the four small corner squares, each having an area of 6¼ units, are added. Thus to "complete the square" one adds $4 \times 6¼ = 25$ units, so obtaining a square of total area $39 + 25 = 64$ units. The side of the larger square must accordingly be 8 units, from which we subtract $2 \times 2½ = 5$ units, to find that $x = 3$.

---

[3] It is of interest to note that the word "algorithm" derives from the author's name. Curiously, "al-jabr" also came to mean "bonesetter" and in sixteenth century Italy the word "algebra" meant the art of restoring or setting broken bones.

[4] The term "root" as a solution to an equation originates with al-Khwarizmi's use of the same word —in its botanical sense— for the unknown in an equation.

Other significant algebraists of the Moslem world include *Abu'l-Wefa* (940–988), *al–Karkhi* (*c.*1029) and the Persian *Omar Khayyam* (*c.*1100), known to posterity as the author of the enchanting *Rubaiyat*. Abul-Wefa gave geometric solutions to some specific quartic equations, and it is believed that al-Karkhi provided the first numerical solutions to such equations as $ax^{2n} + bx^n = c$, in which the Diophantine restriction to rational solutions has been relaxed. In doing so al-Karkhi took the first significant step towards the solution by radicals of algebraic equations of higher than second degree which was to prove of such importance in the later development of algebra.

Omar Khayyam—who regarded algebra as "proven geometric facts"—devoted most of his mathematical efforts to the solution of cubic equations. However, he provided only geometric solutions, mistakenly believing that arithmetic solutions to such equations were, in general, impossible. But he took the significant step of using intersecting conics to solve the general cubic equation (with positive roots), thereby extending the work of Menaechmus and Archimedes. He observes that, since it contains a cube, this equation cannot be solved by plane geometry, that is, by the use of straightedge and comapasses alone. Hampered by the purely rhetorical nature of his algebra, Omar Khayyam had to resort to circumlocutions of painful tortuosity to express his ideas—a far cry from the elegant quatrains of the *Rubaiyat*. In modern notation, his method of solving the cubic went like this. Given a cubic $x^3 + ax^2 + bx + c = 0$, substitute $2py$ for x^2, to obtain $2pxy + 2apy + bx + c = 0$. The latter is the equation of a hyperbola, and the equation $x^2 = 2py$ that of a parabola. The abscissas of the points of intersection of the two curves will be the roots of the given cubic equation.

Algebra in Europe

The first algebraic work of significance to appear in medieval Europe was the *Liber Abaci* (1202) of *Leonardo of Pisa* (*c.*1175–1230), also known as *Fibonacci*. The work's title, which means "Book of the Abacus", is highly misleading, since the use of the abacus is not discussed at all. On the contrary, in it the use of Hindu-Arabic numerals is strongly advocated. It is in the *Liber Abaci* that we first encounter the celebrated *Fibonacci sequence* 1, 1, 2, 3, 5, 8, 13, 21, ..., in which each term after the second is the sum of its two immediate predecessors. Amusingly, this sequence arises as the solution to a problem involving the breeding of rabbits! The sequence has been shown to have many interesting properties, for example, successive terms are relatively prime, and the nth term[5] is given by the expression

$$[(1 + \sqrt{5})^n - (1 - \sqrt{5})^n]/2^n\sqrt{5}.$$

In the *Liber Abaci* and in two later works, the *Liber Quadratorum* and the *Flos* (both 1225), Fibonacci describes, in rhetorical form, solutions of determinate and indeterminate equations of the first and second degree, as well as some cubic equations.

---

[5] I owe to Gerard Khatcherian the observation that the nth term of the Fibonacci sequence may be identified as the total number of ways of ascending an n-stepped staircase taking one or two steps at a time.

Most remarkable is his discussion of the cubic equation $x^3 + 2x^2 + 10x = 20$. He proves that this equation cannot have a "Euclidean" root of the form $a + \sqrt{b}$ with rational a and b. He then states an approximate answer which, expressed in decimal notation, is correct to nine places. This is the most accurate approximation to an irrational root given in Europe up to that time, and exactly how it was obtained remains a mystery to this day. The *Liber Quadratorum* contains solutions to a variety of problems in indeterminate analysis: for example, in find a rational number x such that both $x^2 - 5$ and $x^2 + 5$ are squares of rational numbers, Fibonacci obtains the correct solution 41/12. Fibonacci makes frequent use of the identity

$$(a^2 + b^2)(c^2 + d^2) = (ac + bd)^2 + (bc - ad)^2$$
$$= (ad + bc)^2 + (ac - bd)^2,$$

which had appeared in Diophantus and was also known to the Arabs.

A seminal figure in the development of algebra was the Frenchman *François Viète* (1540–1603). He was probably the first to conceive of algebra in something close to its modern sense as being equally applicable to geometric magnitudes and numbers. For Viète, algebra was a general method of reasoning about "species" or forms of things—for this reason he calls algebra *logistica species* in his *Isagoge in Artem Analyticam* ("Introduction to the Analytic Art") of 1591. In this work, the first genuinely (even if not fully) symbolic algebra, Viète makes a clear-cut distinction between the idea of a *parameter* (e.g., a coefficient in an equation) and that of *unknown quantity*, using consonants to stand for the former and vowels the latter. Viète saw clearly that in studying the general quadratic equation $ax^2 + bx + c = 0$ (in modern notation), one was actually studying an entire class of equations. Viète was also possibly the first to note some of the relations between the roots and coefficients of an equation: for instance, the fact that if the cubic equation $x^3 + b = 3ax$ has two positive roots u and v, then $3a = u^2 + uv + v^2$ and $b = uv^2 + vu^2$.

The Solution of the General Equations of Degrees 3 and 4

A major influence on the development of algebra has been the attempt to formulate general *schemes of solutions for equations*. While the method for solving quadratic equations of "completing the square" was in essence known to the ancient Babylonians, general solutions to cubic and quartic equations did not appear until the sixteenth century in Italy. The general solution to the cubic equation $x^3 + ax^2 + bx + c = 0$ was first discovered around 1500 by *Scipione del Ferro* (1465–1526), and apparently independently in 1530 by *Nicolò Tartaglia* (1499–1557). Tartaglia's solution was actually published by *Girolamo Cardano* (1501–1576) in his *Ars Magna* of 1545 and for this reason is known as *Cardano's solution* (although Cardano, it is true, does credit Tartaglia with the solution).

The *Ars Magna* is largely rhetorical but in modern notation the method of solving the cubic presented in it may be briefly presented as follows. We first remove the quadratic term ax^2 by putting $x = y - \frac{1}{3}a$: this changes the equation into the form

$y^3 + py + q = 0$, where the coefficients p, q are readily expressible in terms of the original coefficients a, b, c. If we now write $y = u + v$, then the equation becomes

$$u^3 + v^3 + (3uv + p)y + q = 0.$$

Thus, if we choose u and v to to satisfy

$$3uv + p = 0,$$

we obtain the two simultaneous equations

$$u^3 + v^3 = -q \qquad u^3v^3 = -p^3/27$$

for the two unknowns u^3 and v^3. By eliminating one of these, we find that the other satisfies the quadratic equation

$$t^2 + qt - p^3/27 = 0.$$

The two roots of this equation are u^3, v^3 and so, using the quadratic formula to write down these roots, we obtain

$$u^3 = -q/2 + \sqrt{(q^2/4 + p^3/27)} \qquad v^3 = -q/2 - \sqrt{(q^2/4 + p^3/27)},$$

whence finally

$$y = u + v = \sqrt[3]{[-q/2 + \sqrt{(q^2/2 + p^3/27)}]} + \sqrt[3]{[-q/2 - \sqrt{(q^2/4 + p^3/27)}]}.$$

An expression involving only the arithmetic operations and the extraction of roots—such as that on the right side of the equation immediately above—is called a *radical*. For this reason it is said that Cardano's procedure shows that the *general cubic equation is soluble by radicals*.

As Cardano recognized, this method of solution leads to trouble in the so-called *irreducible case* when the equation has three real roots (the only other possible number of real roots being 1). To see what happens, consider as an example the equation $x^3 - 2x^2 - x + 2 = 0$. The left hand side factorizes as $(x + 1)(x - 1)(x - 2)$, so the equation has as three real roots -1, 1, 2. Performing the required calculations in Cardano's procedure, we find that $p = -7/3$, $q = 20/27$, so that $q^2/4 + p^3/27 = -972/54^2$ $= -1/3$. But now Cardano's formula tells us to to take the square root of this number, which is negative! Thus, although the roots of the original equation are *known* to be real, Cardano's method will *not* furnish the solution *unless* one is prepared to countenance the use of *complex numbers*. In fact the formula gives (recalling that we write $i = \sqrt{-1}$)

$$y = u + v = \sqrt[3]{(-10/27 + i/3)} + \sqrt[3]{(-10/27 - i/3)}.$$

This expression gives a real result because, while the cube roots in it are all *individually* complex, their *sum* turns out to be real and gives the correct value for[6] y.

It is worth noting in this connection that in 1591 Viète formulated a solution to the cubic in which use is made of a *trigonometric* identity involving the cosine function $\cos A$ of an angle A, so bypassing Cardano's formula, and avoiding altogether the use of complex numbers in the irreducible case. He starts with the well-known identity expressing the cosine of $3A$ in terms of the cosine of A:

$$\cos 3A = 4\cos^3 A - 3\cos A,$$

and writes $z = \cos A$ to obtain

$$z^3 - 3z/4 - \tfrac{1}{4}\cos 3A = 0. \tag{1}$$

If the given cubic is

$$y^3 + py + q = 0, \tag{2}$$

then by substituting $y = hz$ and choosing h appropriately the coefficients of (2) can be made to coincide with those of (1). Making this substitution in (2) gives

$$z^3 + (p/h^2)z + q/h^3 = 0. \tag{3}$$

We require that $p/h^2 = -\tfrac{3}{4}$ so that $h = \sqrt{(-4p/3)}$. Now we select an angle A to satisfy

$$q/h^3 = -\tfrac{1}{4}\cos 3A,$$

that is, to satisfy

$$\cos 3A = -4q/h^3 = -q/(2\sqrt{-p^3/27}) = k. \tag{4}$$

If the three roots of (3) are real then it can be shown that p is negative—so that h is real— and also that $|k| < 1$, so that an angle A can be chosen to satisfy (4). In that case $z = \cos A$ satisfies (1), hence also (3), and so $y = z/h$ satisfies (2).

Viète obtained only the single root $z = \cos A$ but in fact the remaining roots are easily obtained. For observe that, if A is *any* angle satisfying (4), then $z = \cos A$ satisfies (3). But clearly, if A satisfies (4), then so do $A + 120°$ and $A + 240°$. It follows that $z = \cos(A + 120°)$ and $z = \cos(A + 240°)$ are the remaining roots.

---

[6]Actually there are three pairs of values for u and v, namely $(-5 + i\sqrt{3})/6$, $(-5 - i\sqrt{3})/6$; $(1 - 3i\sqrt{3})/6$, $(1 + 3i\sqrt{3})/6$; and $(4 + 2i\sqrt{3})/6$, $(4 - 2\sqrt{3}i)/6$. Summing each pair gives $y = -5/3$ or $1/3$ or $4/3$, whence $x = y + 2/3 = -1$ or 1 or 2.

It is striking that the first genuine use of imaginary or complex numbers was made in the theory of *cubic* equations, and not, as might be supposed, in the theory of quadratic equations, where they are customarily introduced nowadays. In the case of cubic equations it was clear that real solutions *actually existed*, even if presented in a bizarre form, so that the appearance of imaginary quantities could not be avoided, as in the case of quadratic equations, by the mere claim that the equation had no solutions. This was the first step in the process which culminated in the nineteenth century with the complete acceptance of complex numbers

While—as we have seen—Cardano's formula suggests that the sum of complex radicals can give a real result, Cardano himself seems to have remained mystified by the apparent fact and never fully accepted it. The genuineness of the phenomenon was first recognized by Cardano's disciple *Rafaello Bombelli* (c.1526–1573). In Bombelli's *L'Algebra* of 1579 he observes that Cardano's formula applied to the irreducible equation $x^3 - 15x = 4$ yields the solution

$$x = \sqrt[3]{(2 + \sqrt{-121})} + \sqrt[3]{(2 - \sqrt{-121})},$$

while direct substitution shows that $x = 4$ is the sole positive solution. It occurred to him that these apparently conflicting facts could be reconciled if it were the case that

$$\sqrt[3]{(2 + \sqrt{-121})} = 2 + \sqrt{-1}, \quad \sqrt[3]{(2 - \sqrt{-121})} \ 2 - \sqrt{-1}. \tag{5}$$

To this end Bombelli lays down rules for the manipulation of imaginary numbers which are strikingly close to those found in modern expositions. He represents complex numbers as "combinations" of four basis elements, which he terms *piu* "more" (+1), *meno* "less" (–1), *piu di meno* "more than less" (+$\sqrt{-1}$), *meno di meno* "less than less" (–$\sqrt{-1}$). Using his rules, Bombelli proceeds to establish the relations (5): in modern notation, the first of these is verified as follows:

$$(2 + i)^3 = 8 + 3.4i + 3.2.i^2 + i^3 = 8 + 12i - 6 - i = 2 + 11i.$$

Unfortunately, Bombelli's clever observation was of no help in actually producing numerical solutions of irreducible equations, for any attempt to determine algebraically the cube roots of the complex numbers in Cardano's formula invariably leads back to the very cubic whose solution was sought in the first place. For instance, if one tries to determine real a and b so that

$$a + bi = \sqrt[3]{(2 + 11i)}, \quad a - bi = \sqrt[3]{(2 - 11i)} \tag{6}$$

then

$$(a + bi)^3 = 2 + 11i.$$

Expanding the left side and equating real parts gives

$$a^3 - 3ab^2 = 2. \tag{7}$$

But if we multiply the two equations in (6) together, we find

$$a^2 + b^2 = (a + bi)(a - bi) = \sqrt[3]{(4 + 121)} = 5,$$

so that $b^2 = 5 - a^2$. Substituting this value for b^2 into (7) then gives

$$4a^3 - 15a = 2.$$

Setting $x = 2a$ in this yields the equation

$$x^3 - 15x = 4,$$

precisely the equation we started with. This fact justifies the use of the term "irreducible".

After the solving of the general cubic equation it was natural to try to deal similarly with equations of degree 4 (*quartic* or *biquadratic* equations). The general solution to these was discovered by Cardano's pupil *Ludovico Ferrari* (1522–1565), and also appears in the *Ars Magna*, where Cardano writes that the method is "due to Luigi Ferrari, who invented it at my request." The method involves reducing the problem to the solution of cubic and quadratic equations. Here, in modern notation, is one version of it.

Consider the equation

$$x^4 + ax^3 + bx^2 + cx + d = 0.$$

To both sides add $(ex + f)^2$, so obtaining

$$x^4 + ax^3 + (b + e^2)x^2 + (c + 2ef) x + (d + f^2) = (ex + f)^2. \tag{8}$$

Now choose e and f so as to make the left side a perfect square of the form $(x^2 + px + q)^2$. Expanding this and comparing coefficients of powers of x, we obtain

$$2p = a, \quad p^2 + 2q = b + e^2, \quad 2pq = c + 2ef, \quad q^2 = d + f^2.$$

This fixes the value of p. The remaining equations may be written in the form

$$e^2 = p^2 + 2q - b, \quad 4e^2f^2 = (2pq - c)^2, \quad f^2 = q^2 - d.$$

If we substitute the first and third of these into the second, we get

$$(2pq - c)^2 = 4(p^2 + 2q - b)(q^2 - d),$$

that is, using $p = \tfrac{1}{2}a,$

$$(aq - c)^2 = (a^2 + 8q - 4b)(q^2 - d).$$

This is a cubic equation in q and so can be solved for q in terms of a, b, c, d. Once a root has been obtained, the equations above yield the values of e and f. Equation (8) then becomes

$$(x^2 + px + q)^2 = (ex + f)^2,$$

that is,

$$[x^2 + (p + e)x + (q + f)]\,[x^2 + (p - e)x + (q - f)] = 0.$$

Thus we finally get two quadratic equations

$$x^2 + (p + e)x + (q + f) = 0, \qquad x^2 + (p - e)x + (q - f) = 0,$$

whose roots are the four roots of the original equation.

The significance of the discovery of general methods of solving these equations lies not in the details of the procedures themselves, but rather in the fact that these methods show that the roots of any polynomial equation of degree up to 4 can be expressed in terms of the coefficients of the equation using only the operations defined in the field of complex numbers, together with the extraction of roots. This fact, accordingly, was in essence known before 1600.'

The Algebraic Insolubility of the General Equation of Degree Greater than Four

In developing methods for solving polynomial equations it was natural that attention should come to be directed to the question of exactly how many roots an equation possessed. Both Cardano and Descartes asserted that a polynomial equation of degree n has exactly n (real or complex) roots, but gave no proof. In the eighteenth century Euler, D'Alembert and Lagrange each formulated a putative proof, but none of them showed that a root actually existed in the first place. This latter fact—the *Fundamental Theorem of Algebra*—was first given a solid proof by Gauss in 1799. At the same time Gauss showed that an nth degree polynomial can be expressed as a product of linear and quadratic factors with real coefficients—a fact which had been recognized, but not rigorously proved, by his predecessors. An important feature of Gauss's work is that, while it establishes the *existence* of a root, it provides no means of actually *calculating* it. This is probably the first example of a *pure existence proof* in mathematics.

Although prior to Gauss's rigorous proof of the fact it was morally certain that all polynomial equations with real coefficients possessed roots, attempts to obtain roots of equations of degree higher than 4 by algebraic means had met with persistent failure. In 1770 the great French mathematician *Joseph-Louis Lagrange* (1736–1813) undertook a systematic analysis of the methods of solving third and fourth degree equations in the hope that an understanding of exactly why these methods worked might furnish a clue as to how to solve higher degree equations. Lagrange observed that to solve an equation of degree $n = 3$ or 4 one introduces a certain rational function[7] f—a *resolvent* function—of the roots of the equation which is then shown itself to satisfy an equation of lower degree than that of the original; this latter equation—the *reduced* equation—can be solved algebraically and its roots then yield a solution to the given equation. To explain why the reduced equation is of lower degree, Lagrange proved the general result that if a rational function f of the roots of a polynomial equation assumes exactly r values when the roots of the equation are permuted arbitrarily, then f itself is a root of an equation of degree r whose coefficients are rational functions of the coefficients of the given equation. In the cases $n = 3$ and 4 the function f can be chosen to assume exactly $n - 1$ different values under arbitrary permutations of the roots of the given equation, so in both cases the reduced equation is of lower degree.

Lagrange sought to extend this technique to the general quintic (degree 5) equation, and so sought a resolvent function for this case which would satisfy an equation of degree less than 5. But his efforts were in vain, which led him to suspect that the solution of the general equation of degree higher than 4 was impossible.

While Lagrange failed to resolve the problem of the algebraic solubility of general higher-degree equations, his analysis of the problem provided the basis for the work of his successors Abel and Galois, who were finally to establish the algebraic insolubility of the general equation of degree > 4. Moreover, Lagrange's insight that one should consider the values that a rational function assumes when its variables are permuted led to the theory of permutation groups (see below).

Lagrange's work on equations had a direct influence on the Italian mathematician *Paolo Ruffini* (1765–1822), who during the first decade of the nineteenth century made several inconclusive attempts to prove that the general equation of degree > 4 did not admit of algebraic solution. Ruffini did succeed in establishing that no rational function of the n roots of an equation of degree $n > 4$ exists which assumes just 3 or 4 values, so proving that no resolvent function in Lagrange's sense could be found which satisfies an equation of degree < 5.

In 1801 Gauss made an important contribution by showing that every *binomial* equation—that is, of the form $x^n - a = 0$—is algebraically soluble (the cyclotomic equation is the case $a = 1$). Nevertheless, he regarded this as a special case, sharing with Lagrange the suspicion that the general equation would prove to be algebraically insoluble.

The problem was finally settled by the Norwegian mathematician *Niels Henrik Abel* (1802–29). While still in high school he had read Lagrange's and Gauss's—but not, it seems, Ruffini's—work on the theory of equations and at first believed that Gauss's solution of binomial equations could be extended to the general quintic. Soon realizing his error, he then tried to prove that the quintic was algebraically insoluble, an effort which in 1826 was to be crowned with success. It is remarkable that he succeeded in doing this by proving—apparently in total ignorance of Ruffini's work—a result which the latter had assumed without proof, namely, that the roots of an

---

[7] A function is said to be *rational* if it is the quotient of two polynomials.

algebraically soluble equation can always be cast in such a form that each of the radicals in them is a rational function of the roots of the equation together with the roots of unity. The insolubility of the general quintic equation is known today as the *Ruffini-Abel Theorem*.

In his work Abel also discovered a general class of algebraically soluble equations. These, the *Abelian* equations, are those with the property that all their roots are rational functions of any one of them. More precisely, an equation is Abelian if, given any root a, there are rational functions θ_1, ..., θ_{n-1} such that the roots of the equation are given by a, $\theta_1(a)$,..., $\theta_{n-1}(a)$ and, in addition, the θ_i *commute* on a in the sense that $\theta_i(\theta_j(a)) = \theta_j(\theta_i(a))$. The cyclotomic equation $x^n - 1 = 0$ is Abelian in this sense, since, as we have seen on p.40, its roots are representable as powers of any one of them. Other important concepts introduced by Abel are those of a field of numbers and of a polynomial irreducible over a given field. A polynomial is said to be *reducible* over a given field if it can be expressed as the product of two polynomials of lower degree with coefficients in the field; in the opposite case it is *irreducible*. For example, $x^2 + 1$ is irreducible over the real field **R**, but $x^2 - 1$ is reducible, as it can be expressed as the product $(x + 1)(x - 1)$.

Abel's work had left open the question of exactly which equations are, or are not, soluble. This and other questions were answered by *Évariste Galois* (1811–32), whose work—the revolutionary significance of which went quite unrecognized by his contemporaries—proved to be instrumental in laying the foundations for what was later to become known as *abstract algebra*.

In his analysis of algebraic equations Galois took from Lagrange the idea of a *permutation* of the set of roots of an equation. Here we may consider a permutation to be any change in the ordered arrangement of a set of objects (or symbols). For example, the permutation in which the order of the letters a, b, c is changed to c, a, b is written (acb), the notation indicating that each letter is taken into the one immediately following, the first letter being understood to be the successor of the last. Thus in the permutation (acb), the letter a goes to c, c in turn to b and finally b to a. The notation (ac), in which b is omitted, signifies the permutation in which a goes to c, c to a and b remains fixed. If two permutations are performed successively, the resulting permutation is called the *product* or *composite* of the two permutations. Thus the product of (acb) and (ac), written $(acb)(ac)$, is the permutation (bc). Notice that the operation of forming the product of permutations is *not commutative*, that is, it depends on the order in which the permutations are performed. Thus, for example, $(ac)(acb) = (ab)$, while we have seen that $(acb)(ac) = (bc)$. The permutation I which leaves all letters fixed is called the *identity permutation*: clearly the product of I with any permutation leaves the latter unchanged. Notice also that, for any permutation P there is a unique *inverse* permutation whose product with P in either order is the identity permutation I. Thus, for example, since $(acb)(bca) = (bca)(acb) = I$, the inverse of (acb) is (bca). We may sum this up by saying that the collection of all 6 permutations on the set $\{a, b, c\}$, together with its product operation, forms a *group*[8], called the *permutation group* on 3 elements. Similarly, the collection of $n! = 1.2.3....n$

---

[8] The group concept will be formally introduced later in the section on abstract algebra.

permutations of n distinct objects—written S_n—forms the full permutation group on n elements. By a *subgroup* of a group we mean a part of it which contains the product and inverse of any of its elements. (Thus the three permutations (abc), $(abc)^2$ and $(abc)^3 = I$ form a subgroup of the full permutation group S_3 on 3 elements.) A subgroup of the full permutation group on n elements will simply be called a *group of permutations* on n elements.

The problem of solving a polynomial equation by radicals was, as Galois grasped, essentially that of finding radicals whose introduction will make the given equation reducible. For example, the general quadratic equation $x^2 + px + q = 0$ is not reducible but it becomes so on introducing the radical $\sqrt{(\frac{1}{4}p^2 - q)}$, for then $x^2 + px + q$ becomes factorizable as $[x + \frac{1}{2}p + \sqrt{(\frac{1}{4}p^2 - q)}][x + \frac{1}{2}p - \sqrt{(\frac{1}{4}p^2 - q)}]$. Galois associated with each polynomial equation a certain group of permutations of its roots— its *Galois group*—whose properties as a group reflect faithfully the properties of the equation. He showed that, in the case of an equation soluble by radicals, each time an appropriate radical is introduced the associated group of permutations is diminished in such a way as to become a certain kind of subgroup of its predecessor. When all the radicals necessary for solving the equation have been introduced—and the polynomial reduced to a product of linear factors in these radicals—the associated group is reduced to the subgroup consisting of just the identity element. In this way the algebraic solubility, or solution by radicals, of the equation is found to correspond to the existence of a sequence of subgroups of its Galois group satisfying a certain condition (somewhat too involved to be formally introduced here), in which case the Galois group is said to be *solvable*. The algebraic solubility of the equation is thus transformed into the solvability of its Galois group. In the case of the general quintic equation, the Galois group is the full permutation group S_5 on 5 elements, which is not solvable. It follows that the general quintic equation is not algebraically soluble. Certain specific equations, for instance $x^5 - x - 1 = 0$, can also be shown to have S_5 as their Galois group, so they are not algebraically soluble either. On the other hand, since all permutation groups on 2, 3, or 4 elements are solvable, all equations of degrees 2, 3, or 4 are algebraically soluble, as we already know. The Galois group of each of the algebraically soluble equations studied by Abel has the property that products within it are *commutative* (and so solvable), that is, for any permutations P, Q,

$$PQ = QP.$$

A group satisfying this condition is for this reason called *Abelian*.

Galois theory (as it is known) thus provides a complete analysis of the solubility of polynomial equations. It can also be used to demonstrate the impossibility of trisecting a general angle, or of doubling the volume of a cube, by ruler and compass constructions. Given a geometric construction problem, one first sets up an algebraic equation whose solution is the desired quantity: for example, in the case of doubling the cube, the equation is $x^3 - 2 = 0$. The condition for solubility of the construction problem with Euclidean tools is that the equation be algebraically soluble by the *successive extraction of square roots*. In terms of Galois theory a necessary and sufficient condition is that the number of elements of the Galois group of the equation be a power of 2. This can be shown not to be the case for the equations corresponding to the doubling of the cube and the trisection of the angle, so neither of these problems is soluble by means of Euclidean tools[9] (see Appendix 1 for an elementary proof of these facts).

Early Abstract Algebra

Although the group concept is chronologically the first of those we now classify as falling under "abstract algebra", it did not in fact play an explicit role in the early development of the subject, which took place in England and Ireland during the first half of the nineteenth century. The central idea of the algebraists of the "British school"—which included *Sir William Rowan Hamilton* (1805–1865), *Augustus de Morgan* (1806–1871), and *George Boole* (1815–1864)—was to investigate the algebraic laws holding of the various number systems, and of mathematical systems in general, in a purely symbolic manner. In de Morgan's *symbolic algebra,* for example, letters *A, B, C* and symbols of operation + and − are understood as being of an entirely abstract character, and so not necessarily signifying numbers or magnitudes and operations thereupon. Indeed de Morgan claims that "with one exception [that of the equality symbol], no word or sign of arithmetic has one atom of meaning..., the object of which is symbols and their laws of combination, giving a symbolic algebra which may hereafter become the grammar of a hundred distinct significant algebras."

Despite de Morgan's insistence on the arbitrary nature of the signs of his algebra, he seems not to have appreciated that this arbitrariness should also be extended to their *laws of combination.* He was still sufficiently influenced by Kantian philosophy to believe that the basic laws of the algebra of number systems—commutativity, associativity and the like—would apply to any algebraic system whatever. He was proved wrong in this by Hamilton's later invention of the algebra of quaternions (see below).

Important contributions to the development of abstract algebra were made by Boole in his two works, *The Mathematical Analysis of Logic* (1847) and *An Investigation of the Laws of Thought* (1854). Boole's system—known nowadays as *Boolean algebra*—as presented in these two works embodies the first successful application of algebraic methods to *logic.*

Boole seems to have had several interpretations of his algebraic system in mind. In his 1847 work he thinks of each of the basic symbols of his "algebra" as standing for the mental operation of selecting just the objects possessing some given attribute or included in some given class. In 1854 he conceives of these symbols as standing for the attributes or classes themselves. He also recognizes that the algebraic laws he proposes are satisfied if the basic symbols are interpreted as taking just the number values 0 and 1, yielding a system of *binary arithmetic.*

---

[9] It should be pointed out that these two problems, which had plagued mathematicians since antiquity, had been quietly settled in the negative in 1837, before the application of Galois theory, by the French mathematician *Pierre Wantzel* (1814–1848).

Boole used letters $x, y, z,...$ to represent sets of things—numbers, points, lines, ideas, or other entities—selected from a universal set or *universe of discourse* which he denoted by the number 1. For example, if the symbol 1 designates the class of all Canadians, then x might stand for the subclass consisting of all Canadians under twenty-one and y for the subclass of Canadians over six feet tall. The number 0 Boole took to designate the *empty class*, that is, the subclass of the domain of discourse having no members. The addition symbol + was taken by Boole to signify *disjoint union*, so that $x + y$ denotes the class of all objects that are either in x or in y, but not in both. The multiplication symbol . was understood to mean *intersection*, so that $x.y$ (or xy) denoted the class of all objects in both x and y. The sign = was, as usual, taken to indicate identity. The resulting "Boolean" algebra is then a field in which not all of the rules of ordinary algebra hold: for example, in it $1 + 1 = 0$, $x.x = x$ and $x(1 - x) = x$ for arbitrary x.

Boole showed that his algebra provided a simple and effective means of presenting *syllogistic reasoning*. The equation $xy = x$, for example, expresses the assertion that all x's are y's. If now in addition $yz = y$, i.e. if all y's are z's, then we get

$$xz = (xy)z = x(yz) = xy = x,$$

i.e. all x's are z's.

Boole's ideas have since undergone extensive development, and the concept of Boolean algebra now plays a central role in mathematical logic and the design of computer circuits.

CHAPTER 5

THE EVOLUTION OF ALGEBRA, II

Hamilton and Quaternions

IN 1833 HAMILTON PRESENTED a paper before the Irish Academy in which he introduced a formal algebra of real number pairs whose rules of combination are precisely those for complex numbers. The important rule for multiplication of these pairs, corresponding to the rule

$$(a + bi)(c + di) = (ac - bd) + (ad + bc)i$$

is

$$(a,b)(c,d) = (ac - bd, ad + bc),$$

which he interpreted as an operation involving rotation. Hamilton's paper provided the definitive formulation of complex numbers as pairs of real numbers.

Hamilton grasped that his ordered pairs could be thought of as directed entities —*vectors*—in the plane, and he naturally sought to extend this conception to three dimensions by moving from number pairs $(a,b) = a + bi$ to number triples $(a,b,c) = a + bi + cj$. How should the operations of addition and multiplication on these entities then be defined? The operation of addition presented no problem—indeed, the rule is obvious. For several years, however, he was unable to see how to define multiplication for triples. It is part of mathematical legend that, in October 1843, as he was walking with his wife along the Royal Canal near Dublin, in a flash of inspiration he saw that the difficulty would vanish if triples were to be replaced by *quadruples* $(a,b,c,d) = a + bi + cj + dk$ and *the commutative law of multiplication jettisoned.* He had already discerned that for number quadruples one should take $i^2 = j^2 = k^2 = -1$; now in addition he realized that one should not only take $ij = k$, but also $ji = -k$ and similarly $jk = i = -kj$ and $ki = j = -ik$. The remaining laws of operation—associativity, distributivity— were then to be those of ordinary algebra. By the revolutionary act[1] of abandoning the

---

[1] It is worth noting in this connection that Hamilton's mathematical work was strongly influenced by his philosophical views, which were derived in the main from Kant. Kant had maintained that space and time were the two essential forms of sensuous intuition, and Hamilton went so far as to proclaim that, just as geometry is the science of pure space, so algebra must be the science of pure time.

commutative law of multiplication, Hamilton had in effect created a self-consistent new algebra—the *quaternion algebra*. It is said that, so impressed was he by this discovery, with a knife he inscribed the fundamental formula for quaternions $i^2 = j^2 = k^2 = ijk$ on a stone of Brougham Bridge.

Thus a *quaternion* is an expression of the form

$$a + bi + cj + dk = \mathbf{u},$$

where a, b, c, d are real numbers. The terms i, j, k are called *units*. The real number a is called the *scalar part*, and the remainder $bi + cj + dk$ the *vector part*, of **u**. The three coefficients of the vector part of **u** may be thought of as rectangular Cartesian coordinates of a point and the units i, j, k as unit vectors directed along the three axes. Two quaternions $\mathbf{u} = a + bi + cj + dk$ and $\mathbf{u}' = a' + b'i + c'j + d'k$ are *equal* if $a = a'$, $b = b'$, $c = c'$, and $d = d'$. Quaternions are *added* by the rule

$$\mathbf{u} + \mathbf{u}' = (a + a') + (b + b')i + (c + c')j + (d + d')k.$$

The *zero* quaternion is defined as

$$\mathbf{0} = 0 + 0i + 0j + 0k;$$

clearly $\mathbf{u} + \mathbf{0} = \mathbf{u}$ for any quaternion **u**. To each real number a there corresponds a quaternion, namely

$$\mathbf{a} = a + 0i + 0j + 0k.$$

Products of quaternions are calculated using the familiar algebraic rules of manipulation, except that in forming products of the units i, j, k the following rules— mentioned above—are to be observed:

$$jk = i, \quad kj = -i, \quad ki = j, \quad ik = -j, \quad ij = k, \quad ji = -k$$
$$i^2 = j^2 = k^2 = -1$$

Accordingly

$$(a + bi + cj + dk)(a' + b'i + c'j + d'k) = A + Bi + Cj + Dk,$$

where

$$A = aa' - bb' - cc' - dd'$$
$$B = ab' + ba' + cd' - dc'$$
$$C = ac' + ca' + db' - bd'$$

were the two essential forms of sensuous intuition, and Hamilton went so far as to proclaim that, just as geometry is the science of pure space, so algebra must be the science of pure time.

$$D = ad' + da' + bc' - cb'.$$

Thus, for instance,

$$(1 + i + j + k)(1 - i + j + k) = 4k,$$

while

$$(1 - i + j + k)(1 + i + j + k) = 4j,$$

so that *multiplication of quaternions is not commutative.* On the other hand, it may be verified from the above rules for multiplying units that the product of units is associative, and it follows from this that *multiplication of quaternions is associative.*

Quaternions form what is known as a *division algebra*, in that, like the real and complex numbers, division by nonzero elements can be effected. We first determine the *reciprocal* or *inverse* u^{-1} of a nonzero quaternion $u = a + bi + cj + dk$. This will have to satisfy

$$uu^{-1} = u^{-1}u = 1 \tag{1}$$

We note that if we define

$$u^{-1} = a - bi - cj - dk,$$

then

$$uu^{-1} = u^{-1}u = a^2 + b^2 + c^2 + d^2. \tag{2}$$

The quantity on the right side is called the *norm* of u and is written $|u|$. It follows then from (2) that if we define

$$u^{-1} = a/|u| - (b/|u|)i - (c/|u|)j - (d/|u|)k,$$

then (1) will be satisfied.

Now let v be any quaternion. Since multiplication is not commutative, there will in general be two quotients of v by u, namely r for which $ru = v$ and s for which $us = v$. To find these, we multiply $ru = v$ by u^{-1} on both sides on the right to get $r = ruu^{-1} = vu^{-1}$ and $us = v$ on the left to get $s = u^{-1}us = u^{-1}v$.

Since the quaternions form a division algebra, it makes sense to ask whether, like the complex numbers, the *Fundamental Theorem of Algebra* holds for them. This questioned was answered in the affirmative in 1944 when Eilenberg and Niven showed that any polynomial equation $a_0 + a_1x + ... + a_nx^n = 0$ with quaternion coefficients has a quaternion root x.

$$(w + x\mathbf{i} + y\mathbf{j} + z\mathbf{k})(a\mathbf{i} + b\mathbf{j} + c\mathbf{k}) = a'\mathbf{i} + b'\mathbf{j} + c'\mathbf{k}.$$

By multiplying the left side of this equality as quaternions and equating coefficients of both sides we get four equations in the unknowns w, x, y, z whose values are then uniquely determined. From the laws of the multiplication of the units $\mathbf{i}$, $\mathbf{j}$, $\mathbf{k}$, we see that "applying" $\mathbf{i}$ in this sense has the effect of rotating the system of unit vectors $\mathbf{i}$, $\mathbf{j}$, $\mathbf{k}$ about $\mathbf{i}$ so as to bring $\mathbf{j}$ into coincidence with $\mathbf{k}$ and $\mathbf{k}$ with $-\mathbf{j}$; and similarly for "applications" of $\mathbf{j}$, $\mathbf{k}$.

Grassmann's "Calculus of Extension"

In 1844 the German mathematician *Hermann Grassmann* (1809–1877) published a work—*Die Lineale Ausdehnungslehre* ("The Calculus of Extension")—in which is formulated a symbolic algebra far surpassing Hamilton's quaternions in generality. Indeed, Grassmann's algebra is nothing less than a full n-dimensional vector calculus.

Grassmann starts with a set of quantities $\{e_1, ..., e_n\}$ that he calls *primary units*: these may be thought of geometrically as a system of n mutually perpendicular line segments of unit length drawn from a common origin in n-dimensional space. It is assumed that these units can be added together, and multiplied by real numbers. Grassmann's basic notion is that of *extensive quantity*[2], which he defines to be an expression of the form

$$a_1 e_1 + ... + a_n e_n = \mathbf{a},$$

where $a_1, ..., a_n$ are any real numbers. Addition of extensive quantities is defined by

$$\mathbf{a} + \mathbf{b} = (a_1 + b_1)e_1 + ... + (a_n + b_n)e_n.$$

Grassmann introduces a very general concept of *multiplication* of extensive quantities. He first assumes that for each pair e_i, e_j of units a new unit $[e_i e_j]$ called their *product* is given. Then he defines the product of the extensive quantities $\mathbf{a}$ and $\mathbf{b}$ to be the extensive quantity

---

[2] Extensive quantities are to be contrasted with *intensive quantities*. Quantities such as mass or volume are extensive in the sense that they are defined over extended regions of space and are therefore additive: thus two pounds + two pounds = 4 pounds. Vector quantities such as velocity or acceleration, being additive in this sense, also count as extensive quantities. On the other hand, quantities such as temperature or density are intensive in that they are defined at a point and are not additive: thus on mixing two buckets of water each having a uniform temperature of 50 degrees one obtains a quantity of water at a temperature of 50, rather than 100, degrees.

$$[\mathbf{ab}] = a_1b_1[e_1e_1] + a_2b_1[e_2e_1] + \ldots + a_nb_{n-1}[e_ne_{n-1}] + a_nb_n[e_ne_n].$$

By imposing various conditions on the unit products $[e_ie_j]$ Grassmann obtains different sorts of product. For simplicity let us assume that $n = 3$. Then, for example, if we take $[e_2e_1] = [e_1e_2]$, $[e_1e_3] = [e_3e_1]$, and $[e_2e_3] = [e_2e_3]$, the coefficients of $[\mathbf{ab}]$ are

$$a_1b_1, a_2b_2, a_3b_3, \; a_1b_2 + a_2b_1, \; a_1b_3 + a_3b_1, \; a_2b_3 + a_3b_2,$$

and so in this case the laws governing the product are just those of ordinary multiplication. If we agree to take $[e_1e_1] = [e_2e_2] = [e_3e_3] = 0$ and $[e_2e_1] = -[e_1e_2]$, $[e_1e_3] = -[e_3e_1]$, $[e_2e_3] = -[e_3e_2]$, then $[\mathbf{ab}]$ has the coefficients

$$a_2b_3 - a_3b_2, \; a_3b_1 - a_1b_3, \; a_1b_2 - a_2b_1.$$

This Grassmann terms a *combinatory* (also known as *vector*) product. Finally, if we choose $[e_2e_1] = [e_1e_2] = [e_1e_3] = [e_3e_1] = [e_2e_3] = [e_3e_2] = 0$ and $[e_1e_1] = [e_2e_2] = [e_3e_3] = 1$, then $[\mathbf{ab}]$ becomes the numerical quantity

$$a_1b_1 + a_2b_2 + a_3b_3;$$

this Grassmann calls the *inner product* of $\mathbf{a}$ and $\mathbf{b}$.

Grassmann defines the *numerical value*—what we would today call the *norm* or *magnitude*—of an extensive quantity $\mathbf{a} = a_1e_1 + \ldots + a_ne_n$ to be the real number

$$|\mathbf{a}| = (a_1^2 + \ldots + a_n^2)^{\frac{1}{2}}.$$

In the case $n = 3$, if we take e_1, e_2, e_3 to be 3 mutually perpendicular line segments of unit length, then each extensive quantity $\mathbf{a} = a_1e_1 + a_2e_2 + a_3e_3$ is a vector in 3-dimensional space with components a_1, a_2, a_3 along e_1, e_2, e_3. The magnitude $|\mathbf{a}|$ is the length of $\mathbf{a}$, the inner product of $\mathbf{a}$ and $\mathbf{b}$ is the product of the length of $\mathbf{a}$ with that of the projection of $\mathbf{b}$ onto $\mathbf{a}$ (or vice-versa), and the magnitude of the combinatory product $[\mathbf{ab}]$ is the area of the parallelogram with adjacent sides $\mathbf{a}$ and $\mathbf{b}$.

Grassmann also considers the product formed by taking the inner product of an extensive quantity with the combinatory product of two extensive quantities, as in $[\mathbf{ab}]\mathbf{c}$. In the 3-dimensional case this quantity may be interpreted as the volume of a parallelepiped with adjacent sides $\mathbf{a}$, $\mathbf{b}$, $\mathbf{c}$.

The novelty and unconventional presentation of Grassmann's work prevented its full significance from being recognized by his contemporaries and for this reason he was led to publish a revised version of the *Ausdehnungslehre* in 1862. Nevertheless, it proved to be an important influence in the development of the three-dimensional vector calculus whose use was to become standard in physics, and later in mathematics also.

Finite-Dimensional Linear Algebras

Like the complex numbers, quaternions can be regarded as constituting an algebra of *hypercomplex numbers* extending the algebra of the real numbers. The nineteenth century saw the creation of many new algebras of hypercomplex numbers which, while different from the quaternion algebra, continued to share many of its properties.

The British algebraist *Arthur Cayley* (1821–1895) (and independently his friend *J. T. Graves*) formulated an eight unit generalization of the quaternions—the *octonion algebra*. His units were $1, e_1, e_2, ..., e_7$ with multiplication table

$$e_i^2 = -1, \quad e_i e_j = -e_j e_i \quad \text{for } i, j = 1, 2, ..., 7 \text{ and } i \neq j$$
$$e_1 e_2 = e_3, \quad e_1 e_4 = e_5, \quad e_1 e_6 = e_7, \quad e_2 e_5 = e_7, \quad e_2 e_4 = -e_6, \quad e_3 e_4 = e_7, \quad e_3 e_5 = e_6,$$

together with the fourteen equations obtained from the last seven by cyclic permutation of the indices; e.g. $e_2 e_3 = e_1$, $e_3 e_1 = e_2$. An *octonion* is defined to be an expression of the form

$$u_0 + u_1 e_1 + ... + u_7 e_7 = \mathbf{u},$$

where the u_i are real numbers. The *norm* $|\mathbf{u}|$ of $\mathbf{u}$ is defined analogously as for quaternions, viz.,

$$|\mathbf{u}| = u_0^2 + u_1^2 + ... + u_7^2.$$

The product of octonions is obtained termwise as expected:

$$(u_0 + u_1 e_1 + ... + u_7 e_7)(v_0 + v_1 e_1 + ... + v_7 e_7) = u_0 v_0 + ... + u_i v_j \, e_i e_j + ...,$$

where each unit product $e_i e_j$ is calculated according to the above table.

Unlike quaternions, multiplication of octonions is *nonassociative*, that is, in general $\mathbf{u}(\mathbf{vw}) \neq (\mathbf{uv})\mathbf{w}$ (nor is it commutative). Nevertheless, like the quaternions, the octonions form a *division algebra* (a fact which seems to have escaped Cayley) in the sense that any octonion $\mathbf{u} \neq 0$ has a reciprocal $\mathbf{u}^{-1}$ satisfying $\mathbf{u}^{-1}\mathbf{u} = \mathbf{u}\mathbf{u}^{-1} = 1$. In fact we may take

$$\mathbf{u}^{-1} = (u_0 - u_1 e_1 - ... - u_7 e_7)/|\mathbf{u}|.$$

The algebra **C** of complex numbers, **H** of quaternions, and **O** of octonions are all examples of what are known as *linear algebras* over the field **R** of real numbers—as is the algebra **R** itself. This means that the elements of each can be added and multiplied by real numbers and with one another in such a way that the usual algebraic laws—with the possible exception of commutativity and associativity of multiplication—are satisfied. Each is of *finite dimension* in the sense that it is generated by a finite number of "units"—1, 2, 4, and 8 for **R**, **C**, **H**, and **O** respectively. Each is a *division algebra* in the sense that every nonzero element has a reciprocal. Multiplication in **R**, **C** and **H** is associative, and in **R** and **C**, commutative. Once these facts were recognized, mathematicians took the natural step of asking whether any more algebras of this kind remained to be discovered. Gauss was convinced that an extension of the complex numbers preserving all its properties was impossible. When Hamilton abandoned the search for a three-dimensional algebra in favour of the four-dimensional algebra of quaternions he had no proof that a three-dimensional algebra could not exist. And neither did Grassmann.

In the 1860s the American mathematician *Benjamin Peirce*[3] (1809–1880) and his son *Charles Sanders Peirce* (1839–1914) undertook a general investigation of associative linear algebras. The elder Peirce found 162 such algebras, but the only division algebras among them were **R**, **C** and **H**. In 1881 the younger Peirce and the German mathematician *Georg Frobenius* (1849–1917) independently proved that these three are the *only* associative finite dimensional linear division algebras over **R**. The question of whether there are any *non*associative division algebras of this kind other than the octonion algebra **O** was not resolved—in the negative—until 1958, and required the full resources of modern algebraic topology for its resolution.

These algebras have a connection with a problem concerning *sums of squares* considered by *Adolf Hurwitz* (1859–1919) in 1898. Certain identities concerning sums of squares of real (or complex) numbers had been known to mathematicians for some time. Each took the form

$$(\Sigma_1^n x_i^2)(\Sigma_1^n y_i^2) = \Sigma_1^n z_i^2, \tag{1}$$

where the z_i have the form

$$z_i = \Sigma_{j,k=1}^n a_{ijk} x_j x_k$$

with a_{ijk} real numbers. The known ones were identities for $n = 1, 2, 4, 8$. The first of these is the trivial one:

$$x_1^2 y_1^2 = (x_1 y_1)^2.$$

The next two are already nontrivial, namely

$$(x_1^2 + x_2^2)(y_1^2 + y_2^2) = (x_1 y_1 - x_2 y_2)^2 + (x_1 y_2 + x_2 y_1)^2$$

$$(x_1^2 + x_2^2 + x_3^2 + x_4^2)(y_1^2 + y_2^2 + y_3^2 + y_4^2) = z_1^2 + z_2^2 + z_3^2 + z_4^2,$$

---

[3] It was the elder Peirce who, in 1870, formulated the well-known definition: *Mathematics is the science that draws necessary conclusions.*

where

$$z_1 = x_1 y_1 - x_2 y_2 - x_3 y_3 - x_4 y_4$$
$$z_2 = x_1 y_2 + x_2 y_1 + x_3 y_4 - x_4 y_3$$
$$z_3 = x_1 y_3 - x_2 y_4 + x_3 y_1 + x_4 y_2$$
$$z_4 = x_1 y_4 + x_2 y_3 - x_3 y_2 + x_4 y_1.$$

The corresponding identity for $n = 8$ is somewhat tedious to write down. It is not known who first discovered the above identity for $n = 2$. The one for $n = 4$ (which plays a central role in the proof of Lagrange's theorem—mentioned in Chapter 3—that every number is the sum of four squares) seems to have been found by Euler and the one for $n = 8$ by C.F. Degen in 1822.

Each of these identities can be deduced from the properties of multiplication in **C**, **H**, and **O**. For instance, when $n = 2$,

$$
\begin{aligned}
(x_1^2 + x_2^2)(y_1^2 + y_2^2) &= (x_1 + ix_2)(x_1 - ix_2)(y_1 + iy_2)(y_1 - iy_2) \\
&= (x_1 + ix_2)(y_1 + iy_2)(x_1 - ix_2)(y_1 - iy_2) \\
&= [x_1 y_1 - x_2 y_2 + i(x_1 y_2 + x_2 y_1)][x_1 y_1 + x_2 y_2 - i(x_1 y_2 + x_2 y_1)] \\
&= (x_1 y_1 - x_2 y_2)^2 + (x_1 y_2 + x_2 y_1)^2.
\end{aligned}
$$

Matrices

One of the most fruitful algebraic concepts to have emerged in the nineteenth century is that of a *matrix* which was originally introduced by Cayley in 1855 as a convenient notation for linear transformations. A pair of equations of the form

$$x' = ax + by$$
$$y' = cx + dy$$

is said to specify a *linear transformation* of x, y into x', y'. Cayley introduced the array of numbers

$$
\begin{bmatrix}
a & b \\
c & d
\end{bmatrix}
$$

—a 2 × 2 *matrix*— to represent this transformation. In general, the $m \times n$ matrix

$$
\begin{bmatrix}
a_{11} & a_{12} \ldots \ldots a_{1n} \\
a_{21} & a_{22} \ldots \ldots a_{2n} \\
\vdots & \vdots \qquad \vdots \\
\vdots & \vdots \qquad \vdots \\
a_{m1} & a_{m2} \ldots \ldots a_{mn}
\end{bmatrix}
$$

represents the linear transformation of the variables $x_1,..., x_n$ into the variables $x_1',..., x_n'$ given by the equations

$$x_1' = a_{11} x_1 + ... + a_{1n} x_n$$
$$\vdots$$
$$\vdots$$
$$x_m' = a_{m1} x_1 + ... + a_{mn} x_n$$

An $n \times n$ matrix is said to be *square*.

Cayley defined a number of operations on matrices and showed essentially that, for each n, the $n \times n$ matrices form a linear associative algebra over the real numbers. Thus, for $n = 2$, Cayley defined the *sum* of the two matrices

$$\mathbf{M} = \begin{bmatrix} a & b \\ c & d \end{bmatrix} \qquad \mathbf{M}' = \begin{bmatrix} a' & b' \\ c' & d' \end{bmatrix}$$

to be the matrix

$$\mathbf{M} + \mathbf{M}' = \begin{bmatrix} a + a' & b + b' \\ c + c' & d + d' \end{bmatrix}$$

and, for any real number α, the *product* of $\mathbf{M}$ by α to be the matrix

$$\alpha \mathbf{M} = \begin{bmatrix} \alpha a & \alpha b \\ \alpha c & \alpha d \end{bmatrix}$$

To multiply two matrices Cayley analyzed the effect of two successive transformations. Thus if the transformation

$$x' = a_{11} x + a_{12} y$$
$$y' = a_{21} x + a_{22} y$$

is followed by the transformation

$$x'' = b_{11} x' + b_{12} y'$$
$$y'' = b_{21} x' + b_{22} y',$$

then x'', y'' and x, y are connected by the equations

$$x'' = (b_{11} a_{11} + b_{12} a_{21})x + (b_{11} a_{12} + b_{12} a_{22})y$$
$$y'' = (b_{21} a_{11} + b_{22} a_{21})x + (b_{21} a_{12} + b_{22} a_{22})y.$$

Accordingly Cayley defined the product of the matrices to be

$$\begin{bmatrix} b_{11} & b_{12} \\ b_{21} & b_{22} \end{bmatrix} \cdot \begin{bmatrix} a_{11} & a_{12} \\ a_{21} & a_{22} \end{bmatrix} = \begin{bmatrix} b_{11}a_{11} + b_{12}a_{21} & b_{11}a_{12} + b_{12}a_{22} \\ b_{21}a_{21} + b_{22}a_{21} & b_{21}a_{12} + b_{22}a_{22} \end{bmatrix}.$$

Multiplication of matrices is associative but not generally commutative. Cayley observed that—according to his definition—an $m \times n$ matrix can be multiplied only by an $n \times p$ matrix. The *zero* and *unit* $n \times n$ matrices are given by

$$0 = \begin{bmatrix} 0 & .... & 0 \\ 0 & ...... & 0 \\ : & & : \\ : & & : \\ 0 & & 0 \end{bmatrix} \qquad I = \begin{bmatrix} 1 & 0 & .... & 0 \\ 0 & 1 & 0... & 0 \\ 0 & 0 & 1..... & 0 \\ : & & & : \\ 0 & & & ....1 \end{bmatrix}$$

These have the property that for every $n \times n$ matrix **M**,

$$\mathbf{M} + 0 = 0 + \mathbf{M} = \mathbf{M}, \quad 0.\mathbf{M} = \mathbf{M}.0 = 0, \quad \mathbf{M}.\mathbf{I} = \mathbf{I}.\mathbf{M} = \mathbf{M}.$$

The *inverse* $\mathbf{M}^{-1}$—if it exists—of a matrix **M** satisfies $\mathbf{M}\mathbf{M}^{-1} = \mathbf{M}^{-1}\mathbf{M} = \mathbf{I}$. Cayley noted that some matrices fail to have inverses in this sense, so that matrices do not in general form a division algebra. In calculating inverses of matrices Cayley made use of *determinants*, devices originally introduced in the eighteenth century in connection with the solution of systems of linear equations and from which the concept of matrix itself may be considered to have originated. The *determinant* det(**M**) of a 2×2 or 3×3 matrix **M** is given by the following rules: if

$$\mathbf{M} = \begin{bmatrix} a & b \\ a' & b' \end{bmatrix}$$

then

$$\det(\mathbf{M}) = \begin{vmatrix} a & b \\ a' & b' \end{vmatrix} = ab' - ba' \; ;$$

if

$$\mathbf{M} = \begin{bmatrix} a & b & c \\ a' & b' & c' \\ a'' & b'' & c'' \end{bmatrix}$$

then

$$\det(\mathbf{M}) \;=\; \begin{vmatrix} a & b & c \\ a' & b' & c' \\ a'' & b'' & c'' \end{vmatrix} \;=\; a\begin{vmatrix} b' & c' \\ b'' & c'' \end{vmatrix} \;-\; b\begin{vmatrix} a' & c' \\ a'' & c'' \end{vmatrix} \;+\; c\begin{vmatrix} a' & b' \\ a'' & b'' \end{vmatrix}$$

$$= ab'c'' + bc'a'' + ca'b'' - (ac'b'' + ba'c'' + cb'a'').$$

The significance of the determinant in connection with linear equations arises from the fact that a system of equations such as

$$a + bx + cy = 0$$
$$a' + b'x + c'y = 0$$
$$a'' + b''x + c''y = 0$$

has a solution if and only if

$$\begin{vmatrix} a & b & c \\ a' & b' & c' \\ a'' & b'' & c'' \end{vmatrix} \;=\; 0$$

The *minor* corresponding to an entry m in a square matrix $\mathbf{M}$ is the determinant of the matrix obtained from $\mathbf{M}$ by deleting the row and column containing m. Using this idea we can determine the inverse of the 3×3 matrix $\mathbf{M}$ above: it is given by

$$\mathbf{M}^{-1} \;=\; \frac{1}{\det(\mathbf{M})} \begin{bmatrix} A & -A' & A'' \\ -B & B' & -B'' \\ C & -C' & C'' \end{bmatrix}$$

where A is the minor of a, A' that of a', etc. Notice that $\mathbf{M}^{-1}$ exists only when $\det(\mathbf{M}) \neq 0$. When $\det(\mathbf{M}) = 0$, Cayley called $\mathbf{M}$ *indeterminate* (the modern term is *singular*).

So we see that determinants are two-dimensional devices for representing *individual numbers,* and matrices may be considered as having arisen from determinants through a literal reading of their two-dimensional character. Matrices may accordingly be regarded as *two-dimensional numbers.*

An important result concerning matrices announced by Cayley (but not proved fully by him) is now known as the *Cayley-Hamilton theorem* This involves the so-called *characteristic equation* of a matrix, which for a matrix $\mathbf{M}$ is defined to be

$$\det(\mathbf{M} - x\mathbf{I}) = 0.$$

Accordingly if

$$\mathbf{M} = \begin{vmatrix} a & b \\ c & d \end{vmatrix},$$

the characteristic equation is

$$x^2 - (a + d)x + ad - bc = 0.$$

The Cayley-Hamilton theorem is the assertion that, if $\mathbf{M}$ is substituted for x in its characteristic equation, the resulting matrix is the zero matrix. Hamilton had proved the special case for 3×3 matrices; the first complete proof was given by Frobenius in 1878.

The roots of the characteristic equation of a matrix $\mathbf{M}$ turn out to be of particular importance. Consider, for example, the of a 3×3 matrix

$$\mathbf{M} = \begin{bmatrix} a & b & c \\ a' & b' & c' \\ a'' & b'' & c'' \end{bmatrix}.$$

$\mathbf{M}$ represents the linear transformation given by the equations

$$ax + by + cz = u$$
$$a'x + b'y + c'z = v$$
$$a''x + b''y + c''z = w$$

which may in turn be written

$$\begin{bmatrix} a & b & c \\ a' & b' & c' \\ a'' & b'' & c \end{bmatrix} \begin{bmatrix} x \\ y \\ z \end{bmatrix} = \begin{bmatrix} u \\ v \\ w \end{bmatrix}. \tag{1}$$

Here the *column matrices*

$$\mathbf{x} = \begin{bmatrix} x \\ y \\ z \end{bmatrix} \qquad \mathbf{u} = \begin{bmatrix} u \\ v \\ w \end{bmatrix}$$

may be thought of as 3-dimensional vectors with components x, y, z and u, v, w, respectively. Equation (1) may then be written

$$\mathbf{Mx} = \mathbf{u}, \tag{2}$$

which expresses the idea that M *transforms* the vector x into the vector u. Thus a 3×3 matrix can be thought of as a *linear operator* on the space R^3 of 3-dimensional vectors, and in general, an $n \times n$ matrix as a linear operator on the space R^n of n-dimensional vectors.

Ordinarily, the vectors x and Mx do not lie in the same line—are not *collinear*— but an important fact in applications is that, for certain vectors $x \neq 0$, Mx and x are collinear, that is, there exists a real number λ for which

$$Mx = \lambda x. \tag{3}$$

The *characteristic value* or *eigenvalue problem* is the following : given an $n \times n$ matrix M, for which vectors $x \neq 0$ and for what numbers λ does (3) hold? Numbers λ satisfying (3) for some vector x are called *characteristic values* or *eigenvalues* (from German *eigen,* "own") of M.

There is an important connection between the eigenvalues of a matrix and its characteristic equation which is revealed when equation (3) is rewritten in the form

$$(M - \lambda I)x = 0.$$

This equation represents a system of n homogeneous[4] first-degree equations in n unknowns, and it is a basic fact about such a system that it has a nonzero solution precisely when

$$\det(M - \lambda I) = 0.$$

It follows that the eigenvalues of M are precisely the roots of its characteristic equation.

Many natural *geometric* operations in the plane or in space give rise to linear transformations, which in turn may be conveniently represented by their associated matrices. In the plane, for instance, a counterclockwise rotation about the origin through an angle θ:

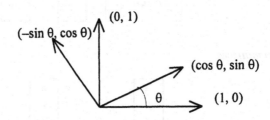

is represented by the matrix

---

[4] A system of first-degree equations in the unknowns $x, y, z, \ldots$ is said to be *homogeneous* if each is of the form $ax + by + cz + \ldots = 0$.

$$\begin{bmatrix} \cos\theta & -\sin\theta \\ \sin\theta & \cos\theta \end{bmatrix}$$

so that, in particular, a 90° rotation is represented by

$$\begin{bmatrix} 0 & -1 \\ 1 & 0 \end{bmatrix}.$$

Reflection in a 45° line, "shear," and projection onto the horizontal axis:

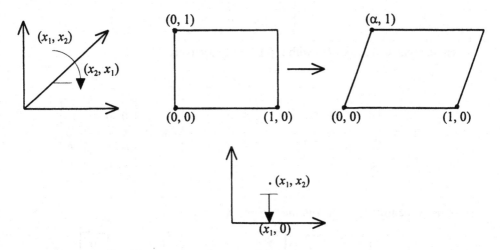

are linear transformations represented by the matrices

$$\begin{bmatrix} 0 & 1 \\ 1 & 0 \end{bmatrix}, \begin{bmatrix} 1 & \alpha \\ 0 & 1 \end{bmatrix}, \begin{bmatrix} 1 & 0 \\ 0 & 0 \end{bmatrix}.$$

Matrices play an important role in representing algebras of hypercomplex numbers. Consider, for instance, the algebra **C** of complex numbers. If with each complex number $a + bi$ we associate the 2×2 matrix

$$\begin{bmatrix} a & -b \\ b & a \end{bmatrix},$$

then it is easily verified that the sum and product of complex numbers corresponds exactly to the sum and product of the associated matrices. In particular the matrix associated with i itself is

$$\begin{bmatrix} 0 & -1 \\ 1 & 0 \end{bmatrix},$$

precisely the matrix corresponding to a 90° counterclockwise rotation.

Quaternions may be represented by *4 × 4 matrices*. In this case, with the quaternion $a + bi + cj + dk$ is associated the matrix

$$\begin{bmatrix} a & -b & -c & -d \\ b & a & -d & c \\ c & d & a & -b \\ d & -c & b & a \end{bmatrix},$$

so that the matrices associated with **i, j, k** are, respectively,

$$\mathbf{i^*} = \begin{bmatrix} 0 & -1 & 0 & 0 \\ 1 & 0 & 0 & 0 \\ 0 & 0 & 0 & -1 \\ 0 & 0 & 1 & 0 \end{bmatrix} \quad \mathbf{j^*} = \begin{bmatrix} 0 & 0 & -1 & 0 \\ 0 & 0 & 0 & 1 \\ 1 & 0 & 0 & 0 \\ 0 & -1 & 0 & 0 \end{bmatrix} \quad \mathbf{k^*} = \begin{bmatrix} 0 & 0 & 0 & -1 \\ 0 & 0 & -1 & 0 \\ 0 & 1 & 0 & 0 \\ 1 & 0 & 0 & 0 \end{bmatrix}.$$

If we now identify **1, i, j, k** with the "column vectors"

$$\mathbf{1} = \begin{bmatrix} 1 \\ 0 \\ 0 \\ 0 \end{bmatrix} \quad \mathbf{i} = \begin{bmatrix} 0 \\ 1 \\ 0 \\ 0 \end{bmatrix} \quad \mathbf{j} = \begin{bmatrix} 0 \\ 0 \\ 1 \\ 0 \end{bmatrix} \quad \mathbf{k} = \begin{bmatrix} 0 \\ 0 \\ 0 \\ 1 \end{bmatrix},$$

then it is readily checked by matrix multiplication that

$$\mathbf{j^*k = i, \quad k^*j = -i, \quad k^*i = j, \quad i^*k = -j, \quad i^*j = k, \quad j^*i = -k,}$$

$$\mathbf{i^*i \;=\; j^*j \;=\; k^*k \;=\; -1.}$$

These are just the multiplication tables for the quaternion units written in terms of the matrices **i*, j*, k*** and the vectors **1, i, j, k**.

It can be shown that *every* linear associative algebra over **R** can be "represented" as an algebra of matrices in this way. To put it succinctly, *hypercomplex numbers are representable as two-dimensional numbers.*

Quaternions can also be represented by matrices built from *complex* numbers. In fact the algebra of quaternions is precisely the same as the algebra generated by the following special matrices with (some) imaginary entries:

$$\sigma_1 = \begin{bmatrix} 0 & i \\ i & 0 \end{bmatrix} \quad \sigma_2 = \begin{bmatrix} 0 & 1 \\ -1 & 0 \end{bmatrix} \quad \sigma_3 = \begin{bmatrix} -i & 0 \\ 0 & i \end{bmatrix} \quad I = \begin{bmatrix} 1 & 0 \\ 0 & 1 \end{bmatrix}$$

where, as usual, $i = \sqrt{-1}$. One easily verifies that under matrix multiplication $\sigma_1, \sigma_2, \sigma_3$, I satisfy precisely the same rules as do the quaternion units $1, i, j, k$. In this representation the general quaternion $a + bi + cj + dk$ is correlated with the complex matrix

$$\begin{bmatrix} a - id & c + ib \\ -c + ib & a + id \end{bmatrix}.$$

These 2×2 complex matrices (and the corresponding quaternions) play an important role in modern quantum physics. There they are regarded as embodying rotations of two dimensional complex vectors known as *spinors*, which describe the state of an electron—a particle with *spin*, that is, intrinsic angular momentum. The matrices σ_1, σ_2 and σ_3, multiplied by $-i$, are the *Pauli spin matrices*. Even the visionary Hamilton, with all his enthusiasm for quaternions, could not have foreseen this application of his invention!

Lie Algebras

While Cayley's algebra of octonions was the first nonassociative linear algebra to be discovered, it is really not much more than a mathematical curiosity. The most significant type of nonassociative algebra—now known as a *Lie algebra*—was introduced in 1876 by the Norwegian mathematician *Sophus Lie* (1842–1899). These algebras were the distillation of Lie's study of the structure of groups of spatial transformations, which in turn involved the idea of an *infinitesimal transformation* in space, that is, one which moves any point in space an infinitesimal distance. Infinitesimal transformations may be added and multiplied by real numbers in a natural way. Given two infinitesimal transformations X, Y, we may also consider their *composite XY*, that is, the transformation that results by applying first Y and then X. Now XY and YX are in general not the same, nor indeed are they infinitesimal transformations. However, the transformation

$$XY - YX$$

is infinitesimal: it is called the *commutator* or *Lie product* of X and Y and is written $[X, Y]$. If we regard this Lie product as a multiplication operation on infinitesimal transformations, then these form a *nonassociative* linear algebra over **R**. Moreover, the Lie product satisfies the laws

(1) $[X, X] = 0$
(2) $[X, Y] = -[Y, X]$
(3) $[[X, Y], Z] + [[Y, Z], X] + [[Z, X], Y] = 0.$

Equation (3), known as the *Jacobi identity*, had already arisen in connection with the researches of the German mathematician *C. J. G. Jacobi* (1804–1851) into the solution of partial differential equations. Equations (1) and (2) are immediate consequences of the definition of $[X, Y]$, and 3) follows by noting that, if in

$$[[X, Y], Z] = XYZ - YXZ - ZXY + ZXY$$

we permute X, Y and Z cyclically, and add all the terms so obtained, we obtain zero.

A linear algebra satisfying (1), (2) and (3) is called a *Lie algebra*. A noteworthy example of a Lie algebra is the algebra of vectors in 3-dimensional space with multiplication given by Grassmann's combinatory or vector product. Any linear associative algebra gives rise to a corresponding Lie algebra whose Lie product is determined by

$$[x, y] = xy - yx.$$

So in particular every algebra of $n \times n$ matrices can be construed as a Lie algebra. Lie himself initially believed he had shown that every Lie algebra can be represented as a Lie algebra of matrices in this sense, but he quickly came to recognize that his proof was incomplete. This fact was established conclusively only in 1935 by the Russian mathematician *I. Ado*.

CHAPTER 6

THE EVOLUTION OF ALGEBRA, III

Algebraic Numbers and Ideals

ONE OF THE MOST SIGNIFICANT PHASES in the development of algebra was the emergence in the nineteenth century of the theory of *algebraic numbers*. This theory grew from attempts to prove Fermat's "Last Theorem" that $x^n + y^n \neq z^n$ for positive integers x, y, z and $n > 2$ (see Chapter 3). The problem was taken up afresh in the eighteen forties by the German mathematician *Ernst Eduard Kummer* (1810–1893). He considered the polynomial $x^p + y^p$ with p prime, and factored it into

$$(x + y)(x + \alpha y)\ldots(x + \alpha^{p-1}y),$$

where α is a complex p^{th} root of unity (see Chapter 3), i.e., a complex solution to the equation

$$x^p - 1 = 0.$$

Since

$$x^p - 1 = (x - 1)(x^{p-1} + x^{p-2} + \ldots + x + 1),$$

and $\alpha \neq 1$, α must satisfy the equation

$$\alpha^{p-1} + \alpha^{p-2} + \ldots + \alpha + 1 = 0. \tag{1}$$

Extending an idea of Gauss, Kummer termed a *complex integer* any number obtained by attaching arbitrary integers to the terms of the expression on the left side of (1), that is, any complex number of the form

$$a_{p-1}\alpha^{p-1} + a_{p-2}\alpha^{p-2} + \ldots + a_1\alpha + a_0,$$

where each a_i is an ordinary integer. The complex integers then form a *ring*, that is, the sum and product of any of them is a number of the same form. This is clearly the case for sums, and the truth of the claim for products follows immediately from the observation that the set of numbers $\{1, \alpha, \dots, \alpha^{p-1}\} = P$ is *closed under multiplication*, that is, the product of any pair of members of P is again a member of P. For suppose that α^m and α^n are members of P with $0 \le m, n \le p - 1$. If $m + n \le p - 1$, then $\alpha^m.\alpha^n = \alpha^{m+n}$ is in P, while if $m + n \ge p$, then, since $\alpha^p = 1$, $\alpha^m.\alpha^n = \alpha^{m+n} = \alpha^p.\alpha^{m+n-p} = \alpha^{m+n-p}$ and the latter is a member of P.

In this connection it is also easily shown that the set P consists of *all* the pth roots of unity. For clearly each member x of P satisfies $x^p - 1 = 0$. Since there are exactly p solutions to this equation, it suffices to show that all the elements of P are distinct. To this end, suppose if possible that $\alpha^m = \alpha^n$ with $0 \le m < n \le p - 1$. Then $\alpha^{n-m} = 1$ and $0 < n - m \le p - 1$. Now let q be the *least* integer such that $\alpha^q = 1$ and $0 < q \le p - 1$. Then since p is prime, p may be written as $sq + r$ with $0 < r < q$, so that $1 = \alpha^p = \alpha^{sq+r} = \alpha^{sq}.\alpha^r = 1. \alpha^r = \alpha^r$. This contradicts the definition of q, and shows that our original assumption that two members of P were identical was incorrect.

Kummer extended to complex integers concepts familiar for the ordinary integers such as primeness, divisibility, and the like, but then mistakenly supposed that the fundamental theorem of arithmetic holds in the ring of complex integers just as it does in the ring of ordinary integers, in other words, that every complex integer factorizes uniquely into primes. It was pointed out by *P.G. Lejeune Dirichlet* (1805–1859) that *this is not always the case for rings of numbers*.

We define a *unit* of a ring to be an element u for which there is an element v such that $uv = 1$; thus a unit is an invertible element of a ring. A factorization $a = a_1 a_2 \dots a_n$ of an element a of a ring is called a *proper factorization* if none of the a_i is a unit or zero. An element of a ring is then said to be *prime* if it has no proper factorization.

Now consider the set of numbers of the form

$$a + b\sqrt{-5},$$

where a, b are integers. It is easily verified that this set is closed under addition and multiplication and so constitutes a ring, which we shall denote by $\mathbf{Z}[\sqrt{-5}]$ to indicate that it is the ring obtained by *adjoining* $\sqrt{-5}$ to the ring $\mathbf{Z}$ of integers. In $\mathbf{Z}[\sqrt{-5}]$ the number 9 can be factorized in two different ways:

$$9 = 3.3 = (2 + \sqrt{-5})(2 - \sqrt{-5}). \tag{2}$$

We claim that these are both *prime* factorizations in $\mathbf{Z}[\sqrt{-5}]$. To establish this, we need first to determine the units of $\mathbf{Z}[\sqrt{-5}]$. Let us define the *norm* of an element $x = a + b\sqrt{-5}$ to be the integer $|x| = a^2 + 5b^2$. It is easily shown that, for any x, y,

$$|xy| = |x||y|. \tag{3}$$

Now if $a + b\sqrt{-5}$ is a unit, then there exists an element $c + d\sqrt{-5}$ such that

$$(a + b\sqrt{-5})(c + d\sqrt{-5}) = 1.$$

Taking the norms of both sides, and using (3), we see that $(a^2 + 5b^2)(c^2 + 5d^2) = |1| = 1$. Since $a^2 + 5b^2$ and $c^2 + 5d^2$ are nonnegative integers, we must have $a^2 + 5b^2 = 1 = c^2 + 5d^2$. It follows that $b = 0$ and $a = \pm 1$, so that $a + b\sqrt{-5} = \pm 1$. Accordingly the only units of $\mathbf{Z}[\sqrt{-5}]$ are ± 1.

Using this fact, we can show that 3 is a prime element of $\mathbf{Z}[\sqrt{-5}]$. Suppose that $3 = (a + b\sqrt{-5})(c + d\sqrt{-5})$. Then, taking the norms of both sides and using the obvious fact that $|3| = 9$, we obtain

$$9 = (a^2 + 5b^2)(c^2 + 5d^2). \tag{4}$$

If neither of the elements $a + b\sqrt{-5}$ and $c + d\sqrt{-5}$ is a unit, their norms must be greater than 1 and it follows from (4) above that each must have norm 3. But it is easily verified that no integers a and b exist for which $a^2 + 5b^2 = 3$, and so no element of $\mathbf{Z}[\sqrt{-5}]$ can have norm 3. Therefore either $a + b\sqrt{-5}$ or $c + d\sqrt{-5}$ must be a unit, and so 3 is prime as claimed.[1]

To show that $2 \pm \sqrt{-5}$ is prime in $\mathbf{Z}[\sqrt{-5}]$, we observe that $|2 \pm \sqrt{-5}| = 9$, so that, were it not prime, it would be factorizable into two elements of norm 3. Since, as we have seen, there are no such elements, it follows that $2 \pm \sqrt{-5}$ is prime. Therefore equation (2) gives two prime factorizations of 9 in $\mathbf{Z}[\sqrt{-5}]$.

So the fundamental theorem of arithmetic fails to hold generallyin number rings of the form $\mathbf{Z}[\sqrt{-n}]$. To remedy this, Kummer adopted the expedient of introducing what he termed *ideal numbers*. In the case of the factorization given by (2), the ideal numbers $\alpha = \sqrt{3}$, $\beta = (2 + \sqrt{-5})/\sqrt{3}$, $\gamma = (1 - \sqrt{-5})/\sqrt{3}$ are introduced, and then 9 can be expressed as the product

$$9 = \alpha^2 \beta \gamma$$

of the "ideal primes" α, β, and γ. (Observe that, in the presence of these ideal numbers, $3 = \alpha^2$ is no longer prime.) In this way the unique factorization into primes is restored.

Although Kummer did in the end succeed in proving Fermat's theorem for all $n \le 100$ by means of his ideal numbers, these were an essentially *ad hoc* device, lacking a systematic foundation. It was Dedekind (whose approach to the foundations of arithmetic was described in Chapter 3) who furnished this foundation, and, in so doing, introduced some of the key concepts of what was to become abstract algebra.

First, Dedekind formulated a general definition of algebraic number which has become standard, and which includes Kummer's complex integers as a special case. An *algebraic number* is a complex number which is a root of an equation of the form

$$a_0 + a_1 x + \ldots + a_n x^n = 0,$$

---

[1] Because 3 is also prime in $\mathbf{Z}$ one might be tempted to suppose that every prime in $\mathbf{Z}$ is also prime in $\mathbf{Z}[\sqrt{-5}]$. That this is not the case can be seen, for example, from the fact that $41 = (6 + \sqrt{-5})(6 - \sqrt{-5})$.

where the a_i are integers. If $a_n = 1$, the roots are called *algebraic integers*. For example, $(-6 + \sqrt{-31})/2$ is an algebraic integer because it is a root of the equation $x^2 + 3x + 10 = 0$. It can be shown that the set of all algebraic numbers forms a *field*, that is, the sum, product, and difference of algebraic numbers, as well as the quotient of an algebraic number by a nonzero algebraic number, are all algebraic numbers. If $u_1,..., u_n$ are algebraic numbers, then the set of algebraic numbers obtained by combining $u_1,..., u_n$ with themselves and with the rational numbers under the four arithmetic operations is clearly also a field: this field will be denoted by $Q(u_1,..., u_n)$ to indicate that it has been obtained by adjoining the algebraic numbers $u_1,..., u_n$ to the field Q of rationals. A field of the form $Q(u_1,..., u_n)$ is called an *algebraic number field*.

The set of algebraic integers in any algebraic number field $Q(u_1,..., u_n)$ includes the ring Z of integers and is closed under all the arithmetical operations with the exception of division and therefore constitutes a *ring*. We denote this ring, which we call a *ring of algebraic integers*, by $Z(u_1,..., u_n)$. The ring $Z(\sqrt{-5})$ of algebraic integers in the field $Q(\sqrt{-5})$ can be shown to coincide with the ring $Z[\sqrt{-5}]$ of numbers of the form $a + b\sqrt{-5}$ discussed above: we have seen that the fundamental theorem of arithmetic fails in this ring, and so does not generally hold in rings of algebraic integers[2].

Dedekind's method of restoring unique prime factorization in rings of algebraic integers was to replace Kummer's ideal numbers by certain *sets* of algebraic integers which, in recognition of Kummer, he called *ideals*, and to show that unique factorization, suitably formulated, held for these.

To understand Dedekind's idea, let us turn our attention to the simplest ring of algebraic integers, the ring Z of ordinary integers. In place of any given integer m, Dedekind considers the set (m) of all *multiples* of m. This set has the two characteristic properties that Dedekind requires of his ideals, namely, if a and b are two members of it, and q is any integer whatsoever, then $a + b$ and qa are both members of it. Moreover, two sets of the form (m) can be *multiplied*: for if we define the product $(m)(n)$ to be the set consisting of all multiples xy with x in (m) and y in (n), then clearly

$$(m)(n) = (mn).$$

In general, if we are given any ring R, an *ideal* in R is a subset I of R which is closed under addition and under multiplication by arbitrary members of R, that is, whenever x and y are in I and r is in R, $x + y$ and rx are both members of I. Any list $\{a_1, ... , a_n\}$ of elements of R *generates* an ideal, denoted by $(a_1, ... , a_n)$, consisting of all elements of R of the form

$$r_1a_1 + ... + r_na_n,$$

---

[2] Nevertheless there do exist rings of the form $Z(\sqrt{-n})$ in which the fundamental theorem of arithmetic holds, namely, when $n = 1, 2, 3, 7, 11, 19, 43, 67, 163$. Numerical evidence had strongly suggested that these were the only possible values of n, and in 1934 *H. A. Heilbronn* (1908–1975) and *E. H. Linfoot* (1905–1982) showed that there could be at most one more such value. Finally, in 1969 *H. M. Stark* proved that this additional value does not exist.

where the r_i are any elements of R. An ideal I is called *principal* if it is generated by a single element a, that is, if $I = (a)$: thus, as before, (a) consists of all the multiples of a. The zero ideal (0) consists of just 0 alone and the unit ideal (1) is, plainly, identical with R itself.

If (a) and (b) are two principal ideals, then it is readily established that (a) is included in (b) exactly when b is a divisor of a, that is, when $a = rb$ for some r. Extending this to ideals, we say that an ideal J is a *divisor* of an ideal I when I is included in J. For principal ideals (a) and (b), the ideal (a, b) is easily seen to be their *greatest common divisor*, in the sense that (a, b) is a divisor of both (a) and (b) and any divisor of (a) and (b) is at the same time a divisor of (a, b).

Ideals may be *multiplied*: if I and J are ideals we define the product IJ to be the ideal consisting of all elements of the form xy, where x and y are elements of I and J respectively. Clearly IJ is included in both I and J. An ideal I is said to be a *factor* of an ideal K if there is an ideal J for which $IJ = K$. If I is a factor of K, then I includes K (but not conversely).

An ideal P is said to be *prime* if—by analogy with integers—whenever P is a divisor of a product IJ of ideals I and J, then P is a divisor of I or of J. It is not difficult to show that this is equivalent to the requirement that whenever a product xy is in P, then at least one of x, y is in P.

The fundamental result of Dedekind's ideal theory is that, *in any ring of algebraic integers, every ideal can be represented uniquely, except for order, as a product of prime ideals.* In particular, every integer u of the ring determines a principal ideal (u) which has such a unique factorization.

Let us illustrate this in the case of the ring $\mathbf{Z}(\sqrt{-5})$ considered above. In this ring the number 21 has the different prime factorizations

$$21 = 3.7 = (1 + 2\sqrt{-5})(1 - 2\sqrt{-5}) = (4 + \sqrt{-5})(4 - \sqrt{-5}) \qquad (5)$$

Now form the greatest common divisor of the principal ideals (3) and $(1 + 2\sqrt{-5})$: this is the ideal $P_1 = (3, 1 + 2\sqrt{-5})$. Consider also the ideals

$$P_2 = (3, 1 - 2\sqrt{-5}), \quad Q_1 = (7, 1 + 2\sqrt{-5}), \quad Q_2 = (7, 1 - 2\sqrt{-5}).$$

Each pair of these ideals may be multiplied simply by multiplying their generating elements, for instance $P_1 P_2 = (9, 3 - 6\sqrt{-5}, 3 + 6\sqrt{-5}, 21)$. The ideal $P_1 P_2$ contains both the numbers 9 and 21, whose g.c.d. is 3. Hence by equation (2) on p. 57, Chapter 4, integers k and ℓ can be found to satisfy $3 = 9k + 21\ell$, and it follows that 3 must be a member of $P_1 P_2$. But all four generators of $P_1 P_2$ are multiples of 3, so $P_1 P_2$ is simply the principal ideal (3) consisting of all multiples of 3 in $\mathbf{Z}(\sqrt{-5})$. Similar computations establish that

$$
\begin{array}{lll}
P_1P_2 = (3), & P_1Q_1 = (1 + 2\sqrt{-5}), & P_1Q_2 = (4 - \sqrt{-5}), \\
Q_1Q_2 = (7), & P_2Q_2 = (1 - 2\sqrt{-5}), & P_2Q_1 = (4 + \sqrt{-5}).
\end{array} \qquad (6)
$$

Each of the ideals P_1, P_2, Q_1, Q_2 may be shown to be prime in $\mathbf{Z}(\sqrt{-5})$. Consider, for instance, P_1. First, we observe that the only (ordinary) integers in P_1 are the multiples of 3. For if a nonmultiple of 3, m say, were in P_1, then, as argued above, the g.c.d. of 3 and m, namely 1, would also be in P_1. In that case there would be elements $a + b\sqrt{-5}$, $c + d\sqrt{-5}$ of $\mathbf{Z}(\sqrt{-5})$ for which

$$1 = 3(a + b\sqrt{-5}) + (1 + 2\sqrt{-5})(c + d\sqrt{-5}).$$

Multiplying out the right side of this equation and equating coefficients of the terms involving, or failing to involve, respectively, $\sqrt{-5}$ gives

$$1 = 3a + c - 10d$$
$$0 = 3b + d + 2a.$$

Multiplying the first of these by 2 and subtracting the second gives

$$6a - 3b - 21d = 2,$$

an impossible relation since the left side is an integer divisible by 3. Therefore the only integers in P_1 are the multiples of 3.

We also observe that, since P_1 is a factor of $P_1 Q_2 = (4 - \sqrt{-5}) = (\sqrt{-5} - 4)$, the number $\sqrt{-5} - 4$ is in P_1. It follows that $\sqrt{-5} - 4 + 3 = \sqrt{-5} - 1$ is also there, and so accordingly is $b\sqrt{-5} - b$ for any integer b. So given any member $u = a + b\sqrt{-5}$ of $\mathbf{Z}(\sqrt{-5})$ we have $u = u' + c$ with $u' = b\sqrt{-5} - b$ in P_1 and $c = a + b$ an integer.

These observations enable us to show that P_1 is prime. For suppose a product uv is in P_1. Then we can find u', v' in P_1 and integers c, d such that $u = u' + c$, $v = v' + d$. Then

$$uv = (u' + c)(v' + d) = u'v' + cv' + du' + cd.$$

But uv, $u'v'$, cv' and du' are all in P_1, and so therefore is cd, which must then, as an integer, be divisible by 3. Thus either c or d is divisible by 3, and so is in P_1. Therefore u or v is in P_1, and so P_1 is prime as claimed. Similar arguments show that P_2, Q_1, and Q_2 are also prime.

The three essentially different factorizations in equation (5) can be regarded as factorizations of the principal ideal generated by 21:

$$(21) = (3)(7) = (1 + 2\sqrt{-5})(1 - 2\sqrt{-5}) = (4 + \sqrt{-5})(4 - \sqrt{-5}).$$

If we now substitute the products given in (6) into this we find that all the factorizations reduce to the same ideal factorization, namely $(21) = P_1 P_2 Q_1 Q_2$. Thus the ideals restore the uniqueness of the factorization.

ABSTRACT ALGEBRA

As we know, algebra began as the art of manipulating *number expressions*, for instance, sums and products of numbers. Since the rules governing such manipulations are the same for all numbers, mathematicians came to realize that the general nature of these rules could best be indicated by employing *letters* to represent numbers. Once this step had been taken, it became apparent that the rules continued to hold good when the letters were interpreted as entities—permutations or attributes, for example—which are not numbers at all. This observation led to the emergence of the general concept of an *algebraic system*, or *structure,* that is, a collection of elements of any sort whatsoever on which are defined operations such as addition or multiplication subject only to the condition of satisfying prescribed rules. The operations of such a structure, together with the rules governing them, may be thought of as *laws of composition,* each of which specifies how two or more elements of the structure are to be *composed* so as to produce a third element of it.

Groups

We have already encountered the algebraic structures called *permutation groups.* As we shall see, the *symmetries* of a given figure, for instance, a square, form a similar type of structure.

On a given plane P fix a point O and a line whose direction we shall call the *horizontal.* Now imagine a rigid square placed on P in such a way that its centre is at O and one side is horizontal. A *symmetry* of the square is defined to be a motion which leaves it looking as it did at the beginning, i.e. with its centre at O and one side horizontal. Clearly the following motions of the square are symmetries:

R: a 90° clockwise rotation about O
R': a 180° clockwise rotation about O
R'': a 270° clockwise rotation about O
H: a 180° rotation about the horizontal axis EF
V: a 180° rotation about the vertical axis GH
D: a 180° rotation about the diagonal 1–3
D': a 180° rotation about the diagonal 2–4

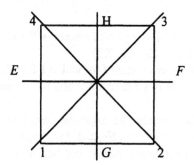

We may *compose* (or *multiply*) two symmetries by performing them in succession. Thus, for example, the *composite* (or *product*) $V.R$ of V and R is obtained by first rotating the square through $180°$ about GH and then through $90°$ about O. By experimenting with a wooden square, one can verify that this has the same total effect as D, rotation about the diagonal 1–3. Another way of checking this is by observing that both $V.R$ and D have the same effect on each vertex of the square: $V.R$ sends 2 into 1 by V and then 1 into 4 by R —hence 2 into 4, just as does D. Similarly, $R.V$ is the result of a clockwise rotation through $90°$, followed by a $180°$ rotation about a vertical axis. It is easily checked that this is the same as D', rotation about the diagonal 2–4. Thus $V.R \neq R.V$, so that composition is *not commutative*. It is, however, easy to see that it is *associative*, i.e., for any symmetries X, Y, Z we have

$$(X.Y).Z = X.(Y.Z).$$

If we compose the two symmetries R and R'' in either order we see that the result is a motion of the square leaving every vertex fixed: this is called the *identity* motion I, which we also regard as a symmetry. Given any symmetry X, it is clear that the motion X^{-1} obtained by *reversing* X is also a symmetry, and that it satisfies

$$X.X^{-1} = X^{-1}.X = I.$$

We call X^{-1} the *inverse* of X.

We shall regard two symmetries as being *identical* if they have the same effect on (the vertices of) the square. Thus, in particular, since the inverse R^{-1} of R has the same effect as R'', we regard them as identical symmetries, and accordingly we may say that R'' is the inverse of R.

We now have a list

$$\{R, R', R'', H, V, D, D', I\}$$

of eight symmetries of the square. This list in fact contains *all* possible symmetries, since any symmetry must carry the vertex 1 into any one of the four possible vertices, and for each such choice there are two possible symmetries, making a total of eight.

The set S of symmetries of the square accordingly has the following properties:

(1) the composite $X.Y$ of any pair of members X,Y of S is a member of S, and for any X, Y, Z in S we have $(X.Y).Z = X.(Y.Z)$;

(2) S contains an *identity* element I for which $I.X = X.I = X$ for all X in S;

(3) for any X in S there is a member X^{-1} of S for which $X.X^{-1} = X^{-1}.X = I$.

These three conditions characterize the algebraic structure known as a *group*. In general, a *group* is defined to be any set S—called the *underlying set* of the group—on which is defined a binary operation (usually, but not always, denoted by ".") called *composition* (or *product*) satisfying the conditions (1) – (3) above. If in addition the composition operation satisfies the *commutative law*, viz.,

$$X.Y = Y.X$$

for any X, Y, the group is called *commutative* or *Abelian*.[3]

We see that the symmetries of a square form a (non-Abelian) group: its *group of symmetries*. In fact, every regular polygon and regular solid (e.g. the cube and regular tetrahedron) has an interesting group of symmetries. For example, the group of symmetries of an equilateral triangle has six elements and is essentially identical with —that is, as we shall later define, *isomorphic to*— the permutation group on its set of three vertices. In general, the group of symmetries of any regular figure may always be regarded as a part (that is, as we shall later define, a *subgroup*) of the permutation group on its set of vertices. This results from the fact that any symmetry of such a figure is uniquely determined by the way it permutes the figure's vertices.

Another important example of a group is the *additive group Z of integers*. The underlying set of this group is the set of all positive and negative integers (together with 0), and the composition operation is *addition*. Notice that in this group 0 plays the role of the identity element. In like fashion we obtain the additive groups of *rational numbers, real numbers*, and *complex numbers*. All of these groups are obviously Abelian.

In forming groups of numbers, as our operation of composition we may use *multiplication* in place of addition, thus obtaining the *multiplicative groups of nonzero*, or *positive, rational numbers; nonzero, or positive, real numbers;* and *nonzero complex numbers*. Notice in this connection that the nonzero integers do not form a multiplicative group since no integer (apart from 1) has a multiplicative inverse which is itself an integer.

Some of the groups we have mentioned are *parts* or *subsets* of others. For example, the additive group of integers is part of the additive group of rational numbers, which is in turn part of the additive group of real numbers. These examples suggest the concept of a subgroup of a group. Thus we define a *subgroup* of a group G to be a set of elements of G (a *subset* of G) which is itself a group under the composition operation of G. It is readily seen that a subset S of a group G is a subgroup exactly when it satisfies the following conditions: (1) the identity element of G is in S, (2) x and y in S imply x. y in S, and (3) x in S implies the inverse x^{-1} is in S. Further examples of subgroups include the subgroup of the group of symmetries of the square consisting of all those symmetries leaving fixed a given vertex, or a given diagonal; and the subgroup of the multiplicative group of rational numbers consisting of all positive real numbers. The reader should have no difficulty in supplying more examples.

Still another important class of groups are the so-called *groups of remainders*. Given an integer $n \geq 2$, the integers $0, 1,..., n - 1$ comprise the possible remainders

---

[3] The term "Abelian" derives from the commuting property of the rational functions associated with Abelian equations: see p. 68 above.

obtained when any integer is divided by n, or, as we say, the remainders *modulo n*. The set

$$H_n = \{0, 1,..., n-1\}$$

can be turned into an additive group by means of the following prescription. For any p, q in H_n we agree that the "sum" of p and q, which we shall write $p \oplus q$, is to be the remainder of the usual sum $p + q$ modulo n. It is then easy to check that H_n, with this operation $\oplus$, is an Abelian group, the *group of remainders modulo n*. In the case $n = 5$, for example, the "addition table" for $\oplus$ is as follows:

$$
\begin{array}{c|ccccc}
\oplus & 0 & 1 & 2 & 3 & 4 \\
\hline
0 & 0 & 1 & 2 & 3 & 4 \\
1 & 1 & 2 & 3 & 4 & 0 \\
2 & 2 & 3 & 4 & 0 & 1 \\
3 & 3 & 4 & 0 & 1 & 2 \\
4 & 4 & 0 & 1 & 2 & 3
\end{array}
$$

In the case $n = 2$ we get the addition table for *binary arithmetic*:

$$
\begin{array}{c|cc}
\oplus & 0 & 1 \\
\hline
0 & 0 & 1 \\
1 & 1 & 0
\end{array}
$$

Since, in H_2, 0 represents the even numbers and 1 the odd numbers, this table is just another way of presenting the familiar rules:

$$even + even = odd + odd = even \quad even + odd = odd.$$

There is a close connection between the additive group of integers and the remainder groups. In order to describe it we need to introduce the idea—one of the most fundamental in mathematics—of a function. A *function*, also called a *transformation, map*, or *correspondence*, between two classes of elements X and Y is any process, or rule, which assigns to each element of X a uniquely determined corresponding element of Y. The class X is called the *domain*, and the class Y the *codomain*, of the function. Using italic letters such as f, g to denote functions, we indicate the fact that a given function has domain X and codomain Y by writing

$$f : X \to Y.,$$

and say that the function f is *from X to Y*, or *defined on X with values in Y*. In this situation, we write $f(x)$ for the element y of Y corresponding under f to a given element x of X: $f(x)$ is called the *image* of x under f, or the *value* of f at x and x is said to be *sent to $f(x)$* by f. An old-fashioned notation for functions is to introduce a function f by writing $y = f(x)$: here x is called the *argument* of f. We shall use this notation occasionally, and especially in Chapter 9.

It is sometimes convenient to specify a function $f: X \to Y$ by indicating its action on elements of its domain X: this is done by writing

$$x \mapsto f(x).$$

It is often helpful to depict a function $f: X \to Y$ by means of a diagram of the sort below, in which arrows are drawn from points in the figure representing the domain of the correspondence to their images in the figure representing its codomain.

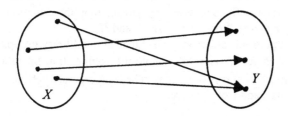

As examples of functions we have

$$x \mapsto x + 1$$
$$x \mapsto \sqrt[3]{x}$$
$$x \mapsto x^2$$

between, respectively, Z and itself, the set of rationals and the set of reals, and the set of reals and the set of non-negative reals. A function is called *one-one* if distinct elements in its domain have distinct images in its codomain, and *onto* if every element of its codomain is the image of at least one element of its domain. A function which is both one-one and onto is called *biunique*, or a *bijection*, or sometimes a *biunique correspondence*. Thus a bijection is a function with the property that each element of its codomain is the image of a unique element of its domain. We see that the first function above is biunique, the second, one-one, and the third, onto.

Functions may also have more than one argument. Thus, for example, we may think of the process which assigns to each triple (x, y, z) of integers the number $x^2 + y^2 + z^2$ as a function f from the set of triples of integers to the set of integers whose value at (x, y, z) is $f(x, y, z) = x^2 + y^2 + z^2$.

Now fix an integer $n \geq 2$. For each integer p in Z write (p) for its remainder modulo n: this yields a function

$$p \mapsto (p)$$

between Z and H_n. This function is obviously onto, but not one-one since it is readily seen that $(p) = (q)$ exactly when $p - q$ is divisible by n. It is now easy to check that we have, for any integers p, q,

$$(p + q) = (p) \oplus (q). \tag{1}$$

For example, taking $n = 5$, $p = 14$, $q = 13$,

$$(14 + 13) = (27) = 2 = 4 \oplus 3 = (14) \oplus (13).$$

Equation (1) tells us that the correspondence $p| \rightarrow (p)$ between Z and H_n transforms the operation $+$ of Z into the operation $\oplus$ of H_n, that is, it *preserves the structure* of the two groups. This fact is briefly expressed by saying that the function $p | \rightarrow (p)$ is a *morphism* (Greek: *morphe*, "form") between Z and H_n.

The concept of morphism may be extended to arbitrary groups. Suppose given two groups G and H: let "." and "$\diamond$" denote the composition operations in G and H respectively. Then a function $f: G \rightarrow H$ is called a *morphism* between G and H if, for any pair of elements x and y of G, we have

$$f(x.y) = f(x) \diamond f(y).$$

The most familiar example of a morphism between groups arises as follows. Let R^p be the multiplicative group of positive real nunbers and let R^a be the additive group of all real numbers. The *common logarithm*[4] $\log_{10} x$ of a positive real number x is defined to be the unique real number y for which

$$10^y = x.$$

The resulting function $x| \rightarrow \log_{10} x$ is then a morphism between R^p and R^a in view of the familiar fact that

$$\log_{10}(xy) = \log_{10} x + \log_{10} y.$$

This function has the further property of being *biunique*. A biunique morphism is called an *isomorphism* (Greek *iso*, "same") and the groups constituting its domain and codomain are then said to be *isomorphic*. Thus $\log_{10}$ is an isomorphism between R^p and R^a; these groups are, accordingly, isomorphic. Isomorphic groups may be regarded as differing only in the notation employed for their elements; apart from that, they may be considered to be identical.

---

[4]Logarithms were introduced by *John Napier* (1550–1617) and developed by *Henry Briggs* (1561–1631). *William Oughtred* (1574–1660) used them in his invention of the *slide rule*, which was to become, by the nineteenth century, the standard calculating instrument cherished by engineers. Only recently has this elegant device—one of the most practical embodiments of the continuous—been superseded by that model of discreteness, the electronic calculator.

Rings and Fields

Another important type of algebraic system is obtained by abstracting at the same time both the additive and the multiplicative structures of the set Z of integers. The result, known as a *ring*, plays a major role in modern algebra. Thus we define a ring (a notion already introduced informally in Chapter 3) to be a system A of elements which, like Z, is an Abelian group under an operation of addition, and on which is defined an operation of multiplication which is associative and distributes over addition. That is, for all elements x, y, z of the ring A,

$$x.(y.z) = (x.y).z \quad x.(y + z) = x.y + x.z \quad (y + z).x = y.x + z.x.$$

If in addition we have always

$$x.y = y.x,$$

then A is called a *commutative ring*.

Naturally, Z with its usual operations of addition and multiplication is then a commutative ring: as such, it is denoted by $\mathbf{Z}$. So, indeed, is the set $\{0, \pm 2, \pm 4, ...\}$ of even integers, as is the set of all multiples of any fixed integer n: this ring is denoted by $\mathbf{Z}_n$. Each of the remainder groups H_n for $n \geq 2$ becomes a commutative ring—called a *remainder ring*—if we define the multiplication operation $\diamond$ by

$$p \diamond q = \text{remainder of } pq \text{ modulo } n.$$

The sets Q, R, and C of rational numbers, real numbers, and complex numbers, respectively, are obviously commutative rings. Earlier in this chapter we also encountered various example of rings of *algebraic numbers*, such as, for each integer n, the ring $\mathbf{Z}[\sqrt{n}]$ consisting of all numbers of the form $a + b\sqrt{n}$, where a, b are integers. As an example of a *noncommutative* ring—that is, one in which multiplication is noncommutative—we have the set of $n \times n$ matrices with real entries for fixed $n \geq 1$: here the operations are matrix addition and multiplication.

Each of the rings Q, R, and C actually has the stronger property that its set of nonzero elements constitutes a *multiplicative group*. A commutative ring satisfying this condition is called a *field* (the concept of field has already been introduced informally in Chapter 3). Thus the rings of rational numbers, real numbers, and complex numbers are all fields: as such, they are denoted by $\mathbf{Q}$, $\mathbf{R}$, and $\mathbf{C}$. It is also easy to show that a remainder ring H_n is a field exactly when n is prime.

Morphisms between rings are similar to morphisms between groups. Thus, given two rings A and B, we define a *morphism* between A and B to be a function $f: A \rightarrow B$ which preserves both addition and multiplication operations in the sense that, for any elements x, y of A, we have

$$f(x + y) = f(x) + f(y) \quad f(x.y) = f(x).f(y).$$

Here the symbols $+$, $.$ on the left side of these equations are to be understood as denoting the addition and multiplication operations in A, while those on the right side the corresponding operations in B. A biunique morphism between rings is called an *isomorphism*.

For example, for any integer $n \geq 2$, the function $p \mapsto (p)$ considered above is a morphism between the rings $\mathbf{Z}$ and H_n. As another example, if m and n are integers ≥ 2 and n is a divisor of m, then the function from H_m to H_n which assigns to each integer $< m$ its remainder modulo n is a morphism of rings. And the function $x + iy \mapsto x - iy$ is an isomorphism of the field of complex numbers with itself (called *conjugation*).

Earlier in this chapter we encountered the notion of an *ideal* in rings of algebraic numbers. In general, given a commutative ring A, an *ideal* in A is a subset I of A such that, for any x, y in I and a in A, $x + y$ and ax are in I. Thus, for example, for any n, the set Z_n of integers divisible by n is an ideal in the ring of integers $\mathbf{Z}$. There is a close connection between ideals and morphisms of rings. Given a morphism of rings $f \colon A \to B$, it is easily verified that the subset of A consisting of those elements a for which $f(a) = 0$—the *kernel* of f—is an ideal in B. For example, the kernel of the above morphism $p \mapsto (p)$ from $\mathbf{Z}$ to H_n is Z_n.

Ordered Sets

We are all familiar with the idea of an *ordering relation*, the most familiar example of which is the ordering of the natural numbers: $1 \leq 2 \leq 3 \leq \dots$. This relation has three characteristic properties:

    reflexivity $p \leq p$
    transitivity $p \leq q$ and $q \leq r$ imply $p \leq r$
    antisymmetry $p \leq q$ and $q \leq p$ imply $p = q$.

It also has the property of

    totality $p \leq q$ or $q \leq p$ for any p, q.

Now there are many examples of binary relations possessing the first three, but not the fourth, of the properties above. Consider, for instance, the relation of *divisibility* on the natural numbers. Clearly this relation is reflexive, transitive, and antisymmetric, but does not possess the property of totality since, for example, neither of the numbers 2 or 3 is a divisor of the other. Another example may be obtained by considering the "southwest" ordering of points on the Cartesian plane (see Chapter 7). Here we say that one point with coordinates (x, y) is *southwest* of a second point with coordinates (u, v) if $x \leq u$ and $y \leq v$, as indicated in the diagram

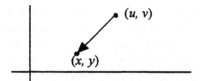

A further example of such a relation is the *inclusion* relation on subsets of a given set. Here we suppose given an arbitrary set U of elements: any set consisting of some (or all) elements of U is called a *subset* of U. We count the *empty set*, denoted by $\varnothing$, possessing no elements at all, as a subset of any set. The collection of all subsets of U is called the *power set* of U and is denoted by Pow(U). Thus, for example, if U consists of the three elements 1, 2, 3, Pow(U) contains the eight different sets $\{1, 2, 3\}$, $\{1, 2\}$, $\{1, 3\}$, $\{2, 3\}$, $\{1\}$, $\{2\}$, $\{3\}$, $\varnothing$. Given two subsets X and Y of U, we say that X is *included* in Y, written $X \subseteq Y$, if every element of X is also an element of Y, as indicated by the diagram

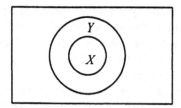

It is now readily verified that the relation of inclusion between subsets of an arbitrary set U is reflexive, transitive, and antisymmetric, but does not possess the totality property if U has more than a single element. Note that inclusion is antisymmetric because two sets with the same elements are necessarily identical.

These examples lead to the following definition. A *partially ordered set* is any set P equipped with a binary relation $\leq$ which is reflexive, transitive, and antisymmetric: in this event the relation $\leq$ is called a *partial ordering* of P. If in addition the relation $\leq$ satisfies the totality principle then P is said to be *totally ordered*. As examples of partially ordered sets we have all those mentioned in the previous paragraph, viz., the natural numbers with the divisibility relation, the points on a plane with the southwest ordering, and the power set of a set with the inclusion relation. Other important examples of partially ordered sets are the collections of all subgroups of a given group, or ideals of a given commutative ring, together with the inclusion relation. As examples of totally ordered sets we have the familiar sets of natural numbers, rational numbers, and real numbers with their usual orderings.

For any partial ordering $\leq$, we write $q \geq p$ for $p \leq q$ and define $p < q$ to mean that $p \leq q$ but $p \neq q$, and we say that q *covers* p if $p < q$ but $p < r < q$ for no r.

Partially ordered sets with a finite number of elements can be conveniently represented by *diagrams*. Each element of the set is represented by a dot so placed that the dot for q is above that for p whenever $p < q$. An ascending line is then drawn from p to q just when q covers p. The relation $p \leq q$ can then be recovered from the diagram by observing that $p < q$ exactly when it is possible to climb from p to q along ascending line segments of the diagram. Here are some examples:

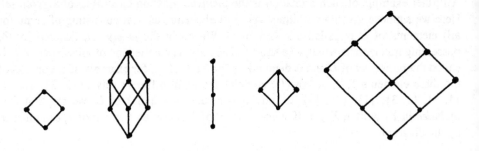

The first of these represents the system of all subsets of a two element set; the second, that of all subsets of a three element set; the third, the numbers 1, 2, 4 under the divisibility relation; the fourth, the numbers 1, 2, 3, 5, 30 under the divisibility relation; the fifth, the nine points (0, 0), (1, 0), (2, 0), (0, 1), (1, 1), (2, 1), (0, 2), (1, 2), (2, 2) in the plane with the southwest ordering.

Just as for groups and rings, we can define the concept of morphism between partially ordered sets. Thus a *morphism* between two partially ordered sets P and Q is a function $f: P \rightarrow Q$ such that, for any p and q in P,

$$p \leq q \text{ implies } f(p) \leq f(q).$$

(Here the left-hand occurrence of the symbol "$\leq$" ignifies the partial ordering of P, and the right-hand occurrence that of Q.) A morphism between partially ordered sets is an *order-preserving function*.

Here are some examples of morphisms between partially ordered sets:

1) the function $x | \rightarrow x^2$ from the totally ordered set of natural numbers to itself;
2) the function $x | \rightarrow \frac{1}{2}x$ from the totally ordered set of rationals to itself;
3) the functions $(x, y) | \rightarrow x$ and $(x, y) | \rightarrow y$ from the set of points on the plane with the southwest ordering to the totally ordered set of real numbers;
4) the function $x | \rightarrow [x]$ from the totally ordered sets of reals to that of integers, where, for each real number x, $[x]$ denotes the greatest integer not exceeding x;
5) the function $X | \rightarrow X^*$ from the power set of the real numbers to that of the integers, where, for each set X of real numbers, X^* denotes the set of integers contained in X;
6) the following functions between finite partially ordered sets indicated by the diagrams

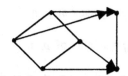 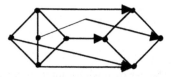

Lattices and Boolean Algebras

Certain of the partially ordered sets we have considered, for example that given by the diagram:

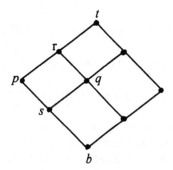

have a special, and important, property. To describe it, consider two typical elements p and q. We see then from the diagram that the element marked r is the least element $\geq$ both p and q in the sense that any element x with this property must satisfy $r \leq x$. Similarly, the element s is the greatest element $\leq$ both p and q in the sense that any element y with this property must satisfy $y \leq s$. The elements r and s are called the *join* and *meet*, respectively, of p and q. It is easy to see that *any* pair of elements of this particular partially ordered set has a join and a meet in this sense.

   This example suggests the following definition. A *lattice* is a partially ordered set P in which, corresponding to each pair of elements p, q there is a unique element, which we denote by $p \vee q$ and call the *join* of p and q, such that, for any element x,

$$p \vee q \leq x \text{ exactly when } p \leq x \text{ and } q \leq x;$$

and also a unique element, which we denote by $p \wedge q$ and call the *meet* of p and q, such that, for any element x,

$$x \leq p \wedge q \text{ exactly when } x \leq p \text{ and } x \leq q.$$

Lattices constitute one of the most important types of partially ordered set. Any totally ordered set is obviously a lattice, in which join and meet are given by:

$$p \vee q = \text{greater of } p, q,$$

$$p \wedge q = \text{smaller of } p, q.$$

The set of positive integers with the divisibility relation is also a lattice in which join and meet are given by

$$p \vee q = \text{least common multiple of } p, q,$$

$$p \wedge q = \text{greatest common divisor of } p, q.$$

The set of points in the plane with the southwest ordering is a lattice, with join and meet given by:

$$(x, y) \vee (u, v) = (max(x, u), \ max(y, v)),$$

$$(x, y) \wedge (u, v) = (min(x, u), \ min(y, v)).$$

For any set U, its power set Pow(U) is also a lattice with join and meet given by

$X \vee Y$ = the set consisting of the members of X together with the members of Y (the set-theoretic *union* of X and Y);
$X \wedge Y$ = the set whose members are those individuals which are in both X and Y (the set-theoretic *intersection* of X and Y).

It is customary to write $X \cup Y$, $X \cap Y$ for the union and intersection of X and Y.

The lattice Pow(U) has several important additional properties that we now describe. First, it satisfies the *distributive laws*, that is, for any X, Y, Z in Pow(U) we have

$$X \cap (Y \cup Z) = (X \cap Y) \cup (X \cap Z),$$

$$X \cup (Y \cap Z) = (X \cup Y) \cap (X \cup Z).$$

These equalities are readily established by the use of elementary logical arguments showing that the sets on either side have the same elements. Secondly, it has *least* and *largest* elements: the empty set $\varnothing$ is least in that it is included in every subset of U, and the set U itself is largest since it includes every subset of U. Finally, Pow(U) is *complemented* in the sense that, for any subset X of U, there is a unique subset X' of U called its (set-theoretic) *complement* such that

$$X \cup X' = U, \quad X \cap X' = \varnothing.$$

X'—which is sometimes written $U - X$—is the subset of U consisting of all elements of U which are not in X.

A lattice satisfying these additional conditions is called a *Boolean algebra*. Thus a *Boolean algebra* is a lattice L which is

(1) *distributive*, that is, for any p, q, r in L,

$$p \wedge (q \vee r) = (p \wedge q) \vee (p \wedge r),$$
$$p \vee (q \wedge r) = (p \vee q) \wedge (p \vee r);$$

(2) *complemented*, that is, possesses elements t ("top") and b ("bottom") such that $b \leq p \leq t$ for all p in P, and corresponding to each p in P there is a (unique) element p' of P for which

$$p \wedge p' = b, \qquad p \vee p' = t.$$

Thus every lattice Pow(U) is a Boolean algebra. On the other hand, for example, the nine element lattice

although distributive, is not a Boolean algebra because its central element has no complement.

The two element lattice

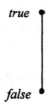

true

false

is obviously a Boolean algebra since it represents the lattice of all subsets of a one element set. It plays a particularly important role in *logic* as it also represents the system of *truth values*. In classical logic propositions are assumed to be capable of assuming just two truth values *true* and *false*. These are conventionally ordered by placing *false* below *true* (so that *false* is "less true" than *true*): the result is the two element Boolean algebra above, which is accordingly known as the *truth-value algebra*.

The four element lattice of all subsets of a two element set is a Boolean algebra which, interestingly, also represents the system of *human blood groups*. Given any species S, define, for individuals s and t of S, $s \leq t$ to mean that t can accept—without ill effects—transfusion of s's blood. Then the relation $\leq$ is clearly reflexive and (presumably) transitive, but not antisymmetric. We call individuals s and t of S *equivalent* if both $s \leq t$ and $t \leq s$: thus equivalent individuals are mutual blood donors. Equivalent individuals are said to be members of the same *blood group*. Accordingly a blood group is the class of individuals equivalent to a given one. In 1901 the Austrian scientist *Karl Landsteiner* (1868–1943) discovered that the human species comprises four such blood groups: these are customarily denoted by *O, A, B*, and *AB*. They form a partially ordered set in a natural way: for any blood groups X and Y we define $X \leq Y$ to mean that $s \leq t$ for any individuals s in X, t in Y. The resulting partially ordered set is the four element Boolean algebra

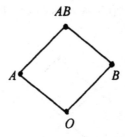

with bottom element *O*—the group of *universal donors*—and top element *AB* —the group of *universal recipients*. It would seem then appropriate to call the four element Boolean algebra the *blood group algebra* or the *Landsteiner algebra*.

The concept of morphism can be defined for lattices and Boolean algebras in the natural way. Thus, a function f is a *morphism* between two lattices L and L' if for any x, y in L we have

$$f(x \vee y) = f(x) \vee f(y), \quad f(x \wedge y) = f(x) \wedge f(y).$$

If L and L' are *Boolean algebras*, f is a *Boolean* morphism if in addition it satisfies, for any element x of L,

$$f(x^*) = f(x)^*,$$

where x^* denotes the complement of x.

It is easy to check that the first two correspondences given in 6), p.104, are lattice morphisms but the third is not (also that the partially ordered sets concerned are lattices). If V is a subset of a set U, then the correspondence $X \mapsto X \cap V$ between Pow(U) and Pow(V) is a Boolean morphism.

The most important fundamental results in the theory of (distributive) lattices and Boolean algebras involve the concept of isomorphism, where two lattices or Boolean algebras are said to be *isomorphic* if there is a biunique morphism between them. Let us define a *lattice of sets* to be a lattice whose members are sets and in which meet and join are set-theoretic intersection and union, respectively, and a *Boolean algebra of sets* to be a Boolean algebra whose members are sets and in which meet, join and complement are set-theoretic intersection, union, and complementation, respectively. Then we have the following *representation theorem* (due to *M. H. Stone*): any distributive lattice, or Boolean algebra, is isomorphic to a lattice, or Boolean algebra, of sets. This result shows that the axioms for distributive lattices or Boolean algebras are an "abstract" characterization of the laws governing the set-theoretic operations, first identified by Boole, of union, intersection, and complementation.

Category Theory

Each of the types of structure—groups, rings, fields, partially ordered sets, lattices, Boolean algebras—we have introduced has an associated notion of morphism, or "structure-preserving" function. In studying these and other mathematical structures, mathematicians came to realize that the idea of morphism between structures was no less important than the idea of structure itself. It was essentially for this reason that *S. Eilenberg* (1914–1998) and *S. Mac Lane* created, in 1945, *category theory*, a general framework for mathematics in which the concepts of mathematical structure and morphism are accorded *equal status*.

In order to motivate the definition of category let us return to groups and morphisms between them. The first and most important fact we note about these morphisms is that they can be *composed*. Thus, suppose we are given two morphisms $f: G \to H$ and $g: H \to J$, such that the codomain H of f coincides with the domain H of G. Then we can define a *composite* morphism $g \circ f : G \to J$ by

$$(g \circ f)(x) = g(f(x))$$

for x in G. It is easy to see that composition in this sense is *associative*, that is, for morphisms $f: G \to H$, $g: H \to J$, $h: J \to K$ we have

$$h \circ (g \circ f) = (h \circ g) \circ f.$$

Another fact is that the identity function $x \mapsto x$ on each group G is a morphism between G and itself: it is called the *identity morphism* on G and written id_G. Clearly identity morphisms have the following property: for any morphism $f: G \to H$,

$$f \circ id_G = id_H \circ f = f.$$

These features provide the basis for the definition of a category. Thus a *category* consists of two collections of entities called *objects* and *arrows*: here objects are to be thought of as the formal counterparts of mathematical structures and arrows as the formal counterparts of morphisms between them. Each arrow f is assigned an object A called its *domain* and an object B called its *codomain*: this fact is indicated by writing $f: A \rightarrow B$. Each object is assigned a morphism $id_A: A \rightarrow A$ called the *identity arrow* on A. For any pair of arrows $f: A \rightarrow B$, $g: B \rightarrow C$ (i.e., such that the domain of g coincides with the codomain of f), there is defined a *composite* morphism $g \circ f: A \rightarrow C$. These prescriptions are subject to the laws of *associativity* and *identity*, viz., given three morphisms $f: A \rightarrow B$, $g: B \rightarrow C$ and $h: C \rightarrow D$ the composites $h \circ (g \circ f)$ and $(h \circ g) \circ f$ coincide; and for any morphism $f: A \rightarrow B$, the composites $f \circ id_A$ and $id_B \circ f$ both coincide with f.

Thus we obtain the *category of groups* whose objects are all groups and whose arrows are all morphisms between them. Similarly, we obtain the categories of *rings, partially ordered sets, lattices,* and *Boolean algebras*: the objects of each of these categories are structures of the sort specified and the arrows morphisms between such structures. Another important category is the *category of sets*, with arbitrary sets as objects and arbitrary functions as arrows.

The fact that any type of mathematical structure determines a corresponding category has led mathematicians to regard category theory as the natural framework for describing the general characteristics of mathematical discourse. By providing a means of expressing the features common to the many branches of mathematics that have appeared in the twentieth century, category theory has come to play an important role in preserving its unity.

CHAPTER 7

THE DEVELOPMENT OF GEOMETRY, I

UNFETTERED BY TRADITION, ALGEBRA made rapid strides during the fifteenth and sixteenth centuries, while geometry, still felt by mathematicians to be in thrall to the towering achievements of the ancient Greeks, languished. But at the beginning of the seventeenth century geometry received a decisive stimulus through the injection of the methods of algebra. This was occasioned largely through the work of Fermat—in his *Ad Locos Planos et Solidos Isagoge,* "Introduction to Plane and Solid Loci", written in 1629 but not published until 1679—and the philosopher-mathematician Descartes—in his *La Géométrie,* which appeared as an appendix to his seminal philosophical work *Discours de la Methode* of 1637. The major effect of the *coordinate*—also known as *algebraic* or *analytic*—geometry they created was to establish a correspondence between curves or surfaces and algebraic equations, thereby opening up geometric investigation to the powerful quantitative methods of the newly emerged algebra.

COORDINATE /ALGEBRAIC/ANALYTIC GEOMETRY

Algebraic Curves

The fundamental idea of coordinate geometry is to associate with each point P in the Euclidean plane a pair of real numbers (x, y)—the *coordinates* of P—giving the distances from two intersecting lines—the *coordinate,* or x- and y-axes—in the plane. The axes are usually taken as being perpendicular—in that case the coordinates are referred to as *rectangular*—and their point of intersection, that is, the point O with coordinates $(0, 0)$, is called the *origin.* The first coordinate x of P is called its *abscissa* (from Latin *abscindere,* "to cut off") and the second coordinate y its *ordinate* (from Latin *ordinare,* "to put in order").

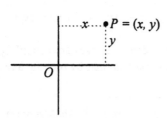

Through the use of coordinates algebraic equations in two unknowns become associated with *curves* in the plane. Given an algebraic equation

$$F(x, y) = 0, \tag{1}$$

where F is a polynomial in x and y with real coefficients, the associated curve—an *algebraic* curve—is obtained by regarding the pair of unknowns as the coordinates of a variable point P, whose position is determined by computing, for each value of x, the corresponding value of y from equation (1). In this way a curve in the plane is traced out:

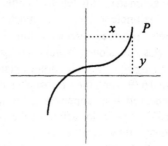

This curve is called the *graph* of—or the curve *represented* by—the equation $F(x, y) = 0$. Descartes took the decisive step of regarding as an admissible geometric object any curve represented by an algebraic equation in this way.

A curve is *classified* by the degree[1] of its representing equation: a curve with a representing equation of degree n is itself said to have degree n. A curve is *irreducible* if its representing equation is irreducible, that is, cannot be factorized into polynomials of lower degree.

Both Descartes and Fermat knew that first-degree curves, with representing equation of the form

$$ax + by + c = 0,$$

are *straight lines*, and that second-degree curves, with representing equation of the form

$$ax^2 + bxy + cy^2 + dx + ey + f = 0, \tag{2}$$

are *conic sections*. Second-degree curves are accordingly also known as *conic* curves. Fermat discovered the general fact that, by rotating axes and translating them parallel to

---

[1] The *degree* of a polynomial equation $F(x, y) = 0$ is the largest sum of the powers of x and y to be found in a term of F. Thus, for example, the equation $x^3y^2 + 3x^2y - 5xy + 2x + 4y + 7 = 0$ has degree 5.

themselves, an equation of the form (2) can always be reduced to one of a number of simpler forms, later shown by Euler to be precisely the following nine:

1. $x^2/a^2 + y^2/b^2 - 1 = 0.$	⬭	An ellipse
2. $x^2/a^2 + y^2/b^2 + 1 = 0.$		An imaginary ellipse
3. $x^2/a^2 + y^2/b^2 = 0.$		A pair of imaginary lines intersecting in a real point
4. $x^2/a^2 - y^2/b^2 - 1 = 0.$	) (	An hyperbola
5. $x^2/a^2 - y^2/b^2 = 0.$	✕	A pair of intersecting lines
6. $y^2 - px = 0.$	⊂	A parabola
7. $x^2 - a^2 = 0.$	∣ ∣	A pair of parallel lines
8. $x^2 + a^2 = 0.$		A pair of imaginary parallel lines
9. $x^2 = 0.$	∣	A pair of coincident straight lines

where none of a, b and p is equal to zero.

Equations 2 and 8 are not satisfied by any real numbers, but only imaginary ones, and accordingly represent *imaginary* curves. The only real number satisfying equation 3 is the pair (0, 0), and so the corresponding curve is reduced to a single point.

The identification of conic sections with equations of degree two was one of the first significant achievements of coordinate geometry.

Cubic Curves

It was *Isaac Newton* (1642–1727), later famed for his theory of universal gravitation, who first undertook the systematic investigation of third-degree, or *cubic* curves. His work on this subject, the *Enumeratio Linearii Tertii Ordinii,* was actually completed by 1676, but not published until 1704. Newton claimed that all curves represented by the general third-degree equation

$$ax^3 + bx^2y + cxy^2 + dy^3 + ex^2 + fxy + gy^2 + hx + jy + k = 0$$

can, by an appropriate choice of coordinate axes, be reduced to one of the following four forms:

$$\text{I.} \qquad xy^2 + ey = ax^3 + bx^2 + cx + d$$
$$\text{II.} \qquad xy = ax^3 + bx^2 + cx + d$$
$$\text{III.} \qquad y^2 = ax^3 + bx^2 + cx + d$$
$$\text{IV.} \qquad y = ax^3 + bx^2 + cx + d.$$

James Stirling (1692–1770) published a proof of Newton's claim in 1717.

Newton also advanced the remarkable claim that *all* cubic curves could be obtained from those of type III by projecting between planes. This assertion was not formally established until 1731, when *Claude-Alexis Clairaut* (1713–1765) and *François Nicole* independently published proofs. Clairaut approached the problem by introducing a surface in three-dimensional space (see section on higher-dimensional spaces below) defined by the equation

$$zy^2 = ax^3 + bx^2z + cxz^2 + dz^3. \qquad (1)$$

This surface is a cubic cone made up of the lines joining the origin to the cubic curve of type III in the plane $z = 1$. Clairaut then established Newton's assertion by showing that every cubic is the intersection of a plane and a cubic cone of the form (1).

By considering the roots of the right-hand sides of equations I – IV, Newton divided cubics into no less than 72 species. Stirling identified four more, and in 1740 another two were discovered, giving a total of 78 types, a huge increase over the nine sorts of conics. Cubics thus form a veritable mathematical zoo.

Returning to the main classification of cubics, I – IV, the most famous curve of type I is the curiously named *Witch of Agnesi* [2], investigated by the Italian mathematician *Maria Gaetana Agnesi* (1718–1799). This curve has equation $xy^2 = a^2(a - x)$ and looks like:

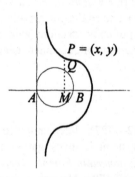

---

[2] The origin of the name "witch" is intriguing. The curve was discussed by Fermat and, in 1718, the Italian mathematician *Luigi Guido Grandi* (1671–1742) gave it the Latin name *versoria*, with the meaning "rope turning a sail," in accordance with its shape. Grandi also supplied the Italian *versiera* for the Latin *versoria*. In her book *Instituzioni Analitiche* of 1748—the first textbook on the calculus to be written by a woman, and a popular book of its day—Agnesi states that the curve is called *la versiera*. In his English translation of Agnesi's book, published in 1801, the British mathematician *John Colson* (1680–1760) apparently mistook "la versiera" for "l'aversiera", meaning "the witch," or "the she-devil." It is with this name that the curve came to be known in the English-speaking world.

A typical point P on the curve has the property that $MQ : MP = AM : AB$, where AQB is a circle of diameter a.

A curve with an equation of type II is called a *Newton's trident* on account of its three-pronged form:

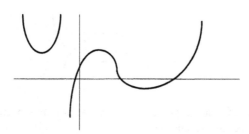

Curves with equations of type III were called by Newton *diverging parabolas*. He discussed the five forms of the curve which arise from the relations among the roots of the cubic equation in x on the right-hand side of III:

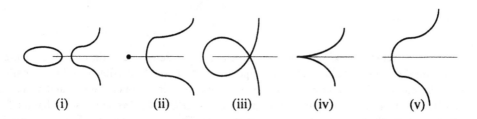

| (i) | (ii) | (iii) | (iv) | (v) |

When all the roots of this equation are real and unequal the curve consists of a closed oval and a parabolic branch, as in the first of the figures immediately above. As the two smaller roots approach each other the oval shrinks to a point and the parabolic branch assumes a more bell-like shape, as in the second figure. If the two greater roots are equal, then the curve assumes the shape shown in the third figure, and is then known as a *nodated parabola*. If all the roots are equal, the curve appears as in the fourth figure, and is then called a *cuspidal* or *semicubical parabola*. Finally, if two roots are imaginary, then the curve has only one bell-like branch, as in the fifth figure.

The most familiar cubic curves—the *cubic parabolas*—are of type IV and look like this:

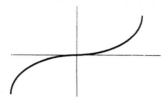

A higher-degree curve such as a cubic can display certain oddities not found in first- or second-degree curves. In particular it may possess *singular points,* that is, points at which the tangent to the curve behaves in an anomalous way. The simplest sort of singular point is an ordinary *double point* or *node,* where two separate branches of the curve cross without touching. Here there is not one, but two tangents, each associated with a branch:

Another type of double point is a *cusp,* at which the two tangents to the separate branches of the curve coincide:

A curve lacking singular points in the sense just described is called *nonsingular.* Nonsingular cubics of type III with rational coefficients—the so-called *elliptic curves*—have come to play an important role in number theory, and were, in particular, instrumental in the resolution of *Fermat's Last Theorem* (see Chapter 3). Here, in a nutshell, is the reason. If Fermat's Last Theorem were false, then there would exist nonzero integers a, b, c and a prime p such that

$$a^p + b^p = c^p.$$

It was observed by the German mathematician Gerhard Frey in 1985 that in that event the associated elliptic curve
$$y^2 = x(x + a^p)(x - b^p)$$

would have some rather unlikely properties. Andrew Wiles showed, in a *tour de force,* that no elliptic curve of this sort could have these properties, and so finally proved Fermat's Last Theorem.

Another feature that a curve may possess (although not classified as a singular point) is a *point of inflection.* At such a point P, the tangent to the curve,

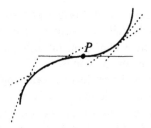

imagined as moving along the curve, comes to rest and reverses its direction of motion. Another way of characterizing a point of inflection is the following. A tangent to a curve at a point may be thought of as a line intersecting the curve in no less than two coincident points; the point is a point of inflection if the tangent there intersects the curve in no less than *three* coincident points. Clairault asserted in 1731 that an irreducible cubic has one, two, or three points of inflection, and *Jean-Paul de Gua de Malves* (1712–1785) proved in 1740 that in the latter case the inflection points must be collinear. Like the vertices of a polygon, singular points and points differ from "typical" points on a curve in possessing exceptional features which assist in characterizing the curve.

In the nineteenth century mathematicians came to realize that the properties of algebraic curves could be considerably simplified if these were to be conceived as possessing "points at infinity", that is, if they were to be regarded as lying in the *projective plane* (see the discussion of projective geometry later in this chapter). The definition of a singular point on a curve in the projective plane is the same as that for a curve in the Euclidean plane. In the projective plane, a point of inflection is called a *flex,* and is defined to be a point on a curve at which the tangent intersects the curve in at least three coincident points. It can be shown that every irreducible cubic in the projective plane possesses either a flex or a singular point.

As an example of an assertion about cubic curves which is true in the projective plane, but not in the Euclidean plane, we may consider the following:

Any straight line intersecting a cubic at least twice intersects it exactly thrice.   (*)

This is false in the Euclidean plane, as can be seen from the graph of the cubic $y^2 = x^3 + x^2 + x$, which looks like

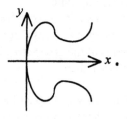

The y-axis is tangent to the curve at the origin and so intersects the curve in two coincident points. But it clearly fails to intersect the curve anywhere else in the Euclidean plane. In the projective plane, on the other hand, the "point at infinity" $(0, 1, 0)$ lies both on the curve and on the y-axis.

The fact that (*) holds in the projective plane enables one to derive the single most striking property of an irreducible cubic curve conceived as lying in that plane, namely, that it has the natural structure of an *Abelian group*. For if P, Q are distinct nonsingular points on an irreducible cubic K, and the line ℓ joining them is not

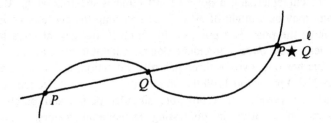

tangent to K at that point, then there is a naturally associated third point $P\bigstar Q$ on K, namely, the third point at which ℓ meets K. This definition can be extended to any two distinct nonsingular points of K by agreeing that $P\bigstar Q = P$ when ℓ is tangent to K at P, and $P\bigstar Q = Q$ when ℓ is tangent to K at Q. When P and Q coincide, and P is not a flex, then ℓ is tangent to K at P and we agree to define $P\bigstar Q$ to be the point where ℓ meets K again. Finally, if P is a flex, we define $P\bigstar Q = P$. In this way we define a binary operation $\bigstar$ on the set K^* of all nonsingular points of K.

Now fix a point O in K^*: O will be called the *base point*. Define a new binary operation $+$ on K^* by

$$P + Q = (P\bigstar Q) \bigstar O.$$

It can be shown that, remarkably, K^* is an Abelian group under the operation $+$, with identity element O, and in which the inverse $-P$ of an element P is given by $-P = P \bigstar (O\bigstar O)$. Moreover, the structure of this group depends only on K, and not on the choice of base point O, for it can, again, be shown that the groups arising from different choices of base point are isomorphic.

When K is irreducible, it must possess a flex, and so it is advantageous to choose one as base point, because then some basic geometric properties of the curve K have elegant interpretations in terms of the group structure. For instance, if for $n \geq 1$ we write nP for $P + P + \ldots + P$ (n times), then

(a) $P + Q + R = O$ if and only if P, Q and R are collinear;
(b) $2P = O$ if and only if the tangent at P passes through O;
(c) $3P = O$ if and only if P is a flex.

A simple consequence of these facts is that, *if P and Q are distinct flexes on an irreducible cubic K, then on the line ℓ joining P and Q there is a third flex of K,* which in turn immediately yields de Gua's theorem that, if a cubic curve has exactly three inflection points, they must be collinear. For choose as base point for K some flex O (which may coincide with P or Q). Then $ℓ$ meets K in a third point R which, by (a) above, satisfies $P + Q + R = O$, so that $R = -P - Q$. Now P and Q are both flexes, so by (c) $3P = O = 3Q$. It follows that $-P = 2P$ and $-Q = 2Q$, whence $R = 2P + 2Q$ and $3R = 6P + 6Q = O + O = O$. So by (c) R is a flex.

Geometric Construction Problems

Descartes' ostensible purpose in introducing algebraic methods into geometry was to provide a general method for solving geometric construction problems. The algebraic method transforms a problem of this type into that of solving a system of equations. A construction then turns out to be performable with Euclidean tools only when the corresponding equations can be solved by applying the usual arithmetic operations $+$, $-$, $\times$ and $\div$ together with the extraction of square roots. The equations associated thereby with the ancient problems of *doubling the cube* and *trisecting the angle* were shown in the nineteenth century to be insoluble without introducing operations of order higher than square roots (actually, cube roots suffice), so that these problems cannot be solved using Euclidean tools alone. (See Appendix 1 for details.)

A further traditional geometric construction problem which, in the formulation made possible by coordinate geometry, exerted a considerable influence on the development of algebra, was that of *constructing a regular polygon of a given number of sides.* The ancient Greeks were able, using Euclidean tools, to construct regular polygons of 3, 4, 5, 6, 8 sides, but failed in their attempts to construct one of 7 sides. As we have seen in Chapter 3, the vertices of a regular n-sided polygon are given, in the complex plane, by the solutions to the cyclotomic equation $z^n = 1$. The solutions to this equation are

$$z = \cos \varphi + i \sin \varphi, \tag{1}$$

where $\varphi = m.360°/n$ for $m = 1, 2, ..., n$. Accordingly the regular n-sided polygon is constructible using Euclidean tools exactly when all the solutions z to (1) are so constructible, which in turn will be the case when the cosine of the angle $360°/n$ is a constructible number (for definition, see Appendix 1). It was shown in the nineteenth century, using the methods of Galois theory, that this is the case just when n has the form

$$n = 2^k. p_1.p_2.p_3....$$

where each p is a so-called *Fermat* prime, that is, a prime number of the form $2^m + 1$, with m itself a power of 2. Since neither 7, nor 9, is of this form, neither the regular heptagon nor the regular nonagon is constructible using Euclidean tools.(See Appendix 1 for an elementary proof of the first assertion.)

Higher Dimensional Spaces.

While coordinate geometry in the plane dominated most of its early development, the suggestion of a coordinate geometry in space can already be found in the works of Fermat and Descartes. In his *Nouveaux Élements des Sections Coniques* of 1679 the French mathematician *Philippe de la Hire* (1640–1718) actually took the explicit step of representing a point in space by three coordinates and writing down the equation of a surface. But the full-scale development of three-dimensional coordinate geometry did not take place until the eighteenth century.

In three-dimensional coordinate geometry three mutually perpendicular axes

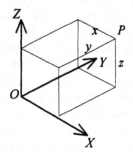

OX, OY and OZ are chosen, and each point P in space is assigned the *coordinates* (x, y, z), where x, y, z are the distances (taken with the corresponding signs) of the point P from the planes OYZ, OXZ and OXY, respectively. The *distance* from the point P to a point Q with coordinates (u, v, w) is then given by

$$d(P, Q) = \sqrt{[(x - u)^2 + (y - v)^2 + (z - w)^2]}. \tag{1}$$

Just as an equation $F(x, y) = 0$ in two variables x, y represents a curve in the plane, so an equation

$$G(x, y, z) = 0$$

represents a *surface* in three dimensional space. A linear equation

$$ax + by + cz + d = 0$$

represents a *plane*. In 1731 Clairault gave the equation

$$x^2/a^2 + y^2/b^2 + z^2/c^2 - 1 = 0$$

for the surface of an ellipsoid and showed that an equation which is homogeneous in x, y and z—that is, all terms of the same degree—represents a cone with vertex at the origin. Not long afterwards *Jacob Hermann* (1678–1733) pointed out that the equation

$$x^2 + y^2 = f(z)$$

represents a surface of revolution about the Z-axis.

The intersection of two surfaces with equations $G(x, y, z) = 0$ and $H(x, y, z) = 0$ represents a *curve* in three-dimensional space. Another way of defining such a curve is to represent it as the trace of the continuous motion of a point. In this case we imagine the coordinates of a point P on the curve as being given by functions $x = x(t)$, $y = y(t)$, $z = z(t)$ of the variable real number t.

The fact that a plane and space can be "coordinatized" by two or three independent coordinates made it inevitable that the question of the geometric interpretation of larger numbers of coordinates would arise. Thus has arisen the conception of "spaces" of any number of dimensions.

In general, for any natural number $n \geq 1$, a *point* in the n–dimensional Euclidean space E_n is an n–tuple of real numbers

$$x = (x_1, ..., x_n);$$

here x_1, ..., x_n are the *coordinates* of x. The *distance* between two points $x = (x_1, ..., x_n)$ and $y = (y_1, ..., y_n)$ is defined by analogy with (1), viz.,

$$d(x, y) = \sqrt{[(x_1 - y_1)^2 + ... + (x_n - y_n)^2]}. \tag{2}$$

For $1 \leq m \leq n$, a set of m equations

$$F_1(x_1, ..., x_n) = 0, ... , F_m(x_1, ..., x_n) = 0$$

determines an $n - m$ dimensional surface in E_n. In particular, $m - 1$ such equations determine a 1-dimensional surface, that is, a curve. Another, more widely used method of specifying a curve —as in 3-dimensional space—is again to represent it as the trace of a continuously moving point. Thus the coordinates of a point on the curve are specified by functions $x_1 = x_1(t)$, ..., $x_n = x_n(t)$ of the variable real number t.

It should be mentioned parenthetically that another, more concrete way of obtaining higher dimensional spaces is to construct them from elements *other than points*. For example, if we consider our ordinary three-dimensional space to be composed of *straight lines* rather than points, that is—to employ the striking metaphor of E. T. Bell—as constituting "a cosmic haystack of long thin straws," as opposed to "an agglomeration of fine birdshot," then it is *four*, as opposed to three, dimensional.

(This is because a line can be specified by its two points of intersection with two fixed planes, and the specification of each of these points requires two coordinates, making four in all.) This gives rise to the so-called *line geometry*. Similarly, *sphere geometry* is obtained by employing *spheres* as primitive elements: this again yields a four-dimensional space since any sphere may be specified by giving its radius and the three spatial coordinates of its centre.

Higher dimensional spaces have proved of especial importance in *physics*, where their use often enables a concept to be presented in a suggestive geometric form. A striking instance of this is furnished by the *dynamical theory of gases*. Suppose that a gas consists of a large number N of molecules. The dynamical state of each of these molecules is then represented by *six* coordinates, three to specify its spatial position, and an additional three giving its three components of velocity. To describe the state of the gas at any given instant we need to specify all the coordinates of the N molecules in the gas. The resulting $6N$ coordinates specify the state of the gas, which is thus represented by a point in a "generalized space" of $6N$ dimensions. A curve in this space —the *state space*—then represents the changing state of the gas through time.

A celebrated application of higher dimensional spaces is to be found in the *four-dimensional spacetime* of *Hermann Minkowski* (1864–1909) which provides the mathematical foundation for *Albert Einstein's* (1879–1955) *special theory of relativity*. The *points* in Minkowski's spacetime are exactly those of four-dimensional Euclidean space E_4, and so each is specified by giving four coordinates (x, y, z, t). But while the coordinates x, y, and z are just the usual ones specifying a position in (three-dimensional) space, the fourth coordinate t is to be understood as indicating a *time*[3]. Accordingly, points in Minkowski's spacetime are correlated, not with "positions" in an abstract *four*-dimensional space, but with *events* occurring at specific *times* and *places* in the familiar *three*-dimensional space of experience. Moreover, the distance between two events—the *interval* between them—is not calculated by means of the usual Euclidean distance formula (2) above, but in accordance with a different formula, one which takes into account the key principles of special relativity that the speed of light is independent of the state of motion of the observer and is an upper limit to all physical velocities. In fact, the interval between two events e and e' with coordinates (x, y, z, t) and (x', y', z', t') is given by

$$d(e, e') = \sqrt{[c^2(t'-t)^2 - [(x'-x)^2 + (y'-y)^2 + (z'-z)^2]]}, \qquad (3)$$

where c is the velocity of light.

To understand the meaning of (3) a little better, imagine that a flash of light emanates from a point P with spatial coordinates (x, y, z). If $x' = x + \Delta x$, $y' = y + \Delta y$, $z' = z + \Delta z$, $t' = t + \Delta t$, then (3) may be put in the form

$$d(e, e')^2 = c^2(\Delta t)^2 - [(\Delta x)^2 + (\Delta y)^2 + (\Delta z)^2]. \qquad (4)$$

---

[3] Thus time constitutes the "fourth dimension" in Minkowski spacetime.

Now $(\Delta x)^2 + (\Delta y)^2 + (\Delta z)^2$ is the squared spatial distance between P and the point P' with coordinates (x', y', z'): this quantity is called the *spatial part* of the (squared) interval between e and e'. The term $c^2(\Delta t)^2$ is the squared distance from P of the wavefront of our flash of light at time $t + \Delta t$: this is called the *temporal part* of the (squared) interval between e and e'. Accordingly the interval between two events is the square root of the *difference* between its temporal and spatial parts. If this difference is *positive*, the interval between or *separation of* the events is called *timelike;* if zero, *lightlike*; and if negative, *spacelike*. A timelike separation of the events e and e' indicates that they could be connected by a physical influence, such as that produced by the motion of a particle. A lightlike separation means that the events can be connected by a light ray. A spacelike separation, on the other hand, would mean that the events can be linked only by an influence travelling faster than light, which, according to the theory of relativity, is impossible. This is reflected in the fact that the interval associated with a spacelike separation would, as the square root of a negative number, be an imaginary quantity.

It is illuminating to map out in a "spacetime diagram" the location of events that can be connected by a light ray to a given event e. For simplicity let us suppose that e takes place at the origin $(0, 0, 0, 0)$ of the spacetime diagram. Then if the spatial coordinates of the event f are (x, y, z), its time coordinate t either has the value

$$t_{\text{future}} = +\sqrt{(x^2 + y^2 + z^2)} \tag{5}$$

or

$$t_{\text{past}} = -\sqrt{(x^2 + y^2 + z^2)} \tag{6}$$

The graphical presentation of this formula is simplified by confining attention to events f whose z-coordinate is zero. In that case the spacetime diagram may be presented as if it had just two spatial coordinates x and y together with the time coordinate t:

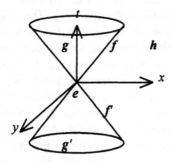

Each event in this diagram with lightlike separation from e either lies, like f, on the surface of the *future light cone* of e (equation 5) or, like f', on the surface of the *past light cone* of e (equation 6). An event such as g contained within the future light cone can be caused by e, and one such as g' can be the cause of e. On the other hand, an event such as *h* which is entirely outside *e*'s light cone cannot be causally related in any way to *e*: it is *causally independent* of *e*.

In this way coordinate geometry has made possible a remarkable fusion of the concepts of space and time. Through the use of imaginary numbers, Minkowski pushed this fusion to its limit. He introduced a new quantity *w* to measure time, defined by

$$w = ict,$$

or

$$\Delta w = ic\Delta t.$$

In that case the expression (4) for the squared interval becomes

$$d(e, e')^2 = (\Delta x)^2 + (\Delta y)^2 + (\Delta z)^2 + (\Delta w)^2,$$

which is precisely the expression for the squared distance in 4-dimensional Euclidean space. Minkowski was sufficiently impressed by this fact to write the famous words:

> Henceforth space by itself, and time by itself, are doomed to fade away into mere shadows, and only a kind of union of the two will have an independent reality.

NONEUCLIDEAN GEOMETRY

Euclid's postulates were put forward as a body of assertions about lines and points conceived as lying in physical space whose correctness would be immediately obvious to everyone. There is one postulate, however, whose correctness is by no means obvious. This is the *fifth*—or *parallel*—*postulate* which, in Playfair's (*John Playfair, 1748–1819*) formulation, states that through any point not on a given line *one and only one line* may be drawn parallel to the given line. The striking feature of this postulate is that it makes an assertion about the *whole extent* of a straight line, imagined as being extended indefinitely in either direction; for two lines are *defined* to be parallel if they never intersect, however far they are produced. Now of course there are many lines through a point which do not intersect a given line within any prescribed finite distance, however large. Since the maximum possible length of an *actual* ruler, thread, or even light ray visible through a telescope is certainly finite, and since within any finite circle there are infinitely many straight lines passing through a given point and not intersecting a given straight line, it follows that the postulate can never be verified —or even refuted—by experiment. On the other hand, all the other postulates of Euclidean geometry have a *finite* character in that they deal with bounded portions of lines and planes. The fact that the parallel postulate is not experimentally verifiable, while the remaining postulates are, suggested the idea of trying to *derive* it from the latter. For centuries, mathematicians strove without success to find such a derivation.

One of the first attempts in this direction was made by *Proclus* (4th century B.C.), who tried to dispense with the need for a special parallel postulate by the ingenious expedient of *defining* the parallel to a given line to be the locus of all points at a fixed distance from the line. Unfortunately, it then became necessary to show that the locus of such points is indeed a straight line! Since this assertion is actually *equivalent* to the parallel postulate, Proclus made no real advance here.

Not until 1733 was the first truly scientific investigation of the parallel postulate published. In that year there appeared the book *Euclides ab onmi naevo vindicatis*—"Euclid Freed of Every Flaw"—by the Italian Jesuit *Girolamo Saccheri* (1667–1733). Without using the parallel postulate, Saccheri easily showed that if, in a quadrilateral *ABCD*, the angles at *A* and *B* are right angles and sides *AD* and *BC* are equal, then the angles at *D* and *C* are also equal. There are then three possibilities: the angles at *C* and *D* are equal *acute, right, or obtuse* angles. Saccheri showed, *assuming that straight lines are indefinitely extensible,* that the case of obtuse angles is

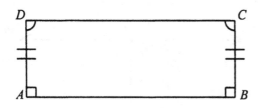

impossible (using, of course, the remaining postulates of Euclidean geometry). His aim was then to demonstrate that the acute angle hypothesis also leads to a contradiction, thus leaving the right angle case, which is easily shown to be equivalent to the parallel postulate. Saccheri's method was thus to assume the acute angle hypothesis, together with the postulates of Euclidean geometry apart from the parallel postulate and attempt to derive a contradiction, a successful outcome showing the parallel postulate to be a consequence of the remaining ones. Remarkably, in the course of his investigations, Saccheri derived many of the theorems of what was later to become known as *noneuclidean geometry*. Unfortunately, however, he completed his discussion by deriving an unconvincing "contradiction" involving a nebulous idea of "infinite element". Had he not been so eager to exhibit a contradiction here, but had rather admitted his inability to find one, Saccheri would today indisputably be credited with the discovery of noneuclidean geometry. Nevertheless, despite its lack of boldness, Saccheri's work, along with that of *Johann Heinrich Lambert* (1728–1777) and *Adrien-Marie Legendre* (1752–1833), suggested the possiblity of a *new* geometry in which the parallel postulate was no longer affirmed.

At that time, any geometric system not absolutely in accordance with Euclid's would have been regarded as nonsensical. However, their continual failure to find a proof of the parallel postulate finally convinced mathematicians that it must be truly *independent* of the others, and that therefore a self-consistent *noneuclidean* geometry is conceivable.

Janos Bolyai (1802–1860) and *Nikolai Ivanovich Lobachevsky* (1793–1856) were the first to publish, independently, in 1832 and 1829, respectively, detailed accounts of a system of noneuclidean geometry[4]. This *Bolyai-Lobachevsky—*also known as *hyperbolic— geometry* possesses certain curious features that set it apart from Euclidean geometry in a most dramatic way, and which fully justify Bolyai's description of it as a "strange new universe". To begin with, there is always more than one straight line parallel to a given one passing through a given point outside it. Let us examine this situation a little more closely. Calling the given line ℓ and the point P, drop perpendicular PQ to ℓ and let m be the perpendicular through P to PQ. Consider one ray PS of m and various rays between PS and PQ. Some of these rays, such as PR, will intersect ℓ, while others, such as PY, will not. As R recedes indefinitely on ℓ from Q, PR will approach a certain *limiting* ray PX that does *not* meet ℓ. The ray PX is "limiting" in the sense that any ray between PX and PQ intersects ℓ, whereas any other ray PY such that PX is between PY and PQ will fail to do so. The ray PX is called the

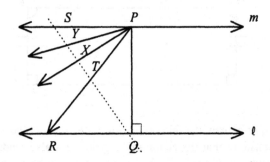

left limiting parallel ray to ℓ through P. Similarly, there is a *right* limiting parallel ray PX' on the opposite side of PQ. These limiting rays are symmetrically situated

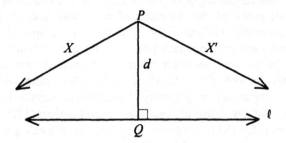

about PQ. The angle $QPX = QPX'$ is called the *angle of parallelism* at P with respect to ℓ: the size of this angle depends only on the distance d from P to Q, and not on the

---

[4] Both were in fact anticipated by Gauss, who, however, fearing critical reaction—to which he referred as "the cries of the Boeotians"—, never published his discoveries in noneuclidean geometry.

identity of the particular line ℓ, nor on that of the particular point P. As the distance d varies, the angle of parallelism α takes on all values between 0 and 90 degrees: as P approaches Q, α approaches 90°, and as P recedes to infinity from Q, α approaches 0°.

Triangles also behave oddly in Bolyai-Lobachevsky geometry. All triangles have angle sum less than 180°, so that, if we define the *defect* of a triangle to be the difference between 180° and the sum of its angles, the defect of any triangle is always positive. As a triangle shrinks, however, its defect becomes arbitrarily small. Another striking fact about triangles in Bolyai-Lobachevsky geometry is that *all similar triangles are congruent.* This means that a given triangle cannot be enlarged or shrunk without changing its shape. A startling consequence is that a segment can be determined with the aid of an angle: for example, an angle (50°, say) of an equilateral triangle determines the length of the side uniquely. To put it more dramatically, Bolyai-Lobachevsky geometry has an *absolute measure of length.* This contrasts starkly with Euclidean geometry. For while it shares with Bolyai-Lobachevsky geometry the feature of possessing an absolute measure of angle in the form of the right angle, it cannot possess an absolute measure of length since within it the geometric properties of figures are invariant under change of scale.

A further curious feature of Bolyai-Lobachevsky geometry is that all convex quadrilaterals have angle sum less than 360°: in particular, there are no quadrilaterals containing four right angles, that is, *there are no rectangles or squares.* Since the customary system of measuring *area* is based on square units, this makes the task of defining area a somewhat ticklish affair. In fact the only reasonable way of defining the area of a triangle is to make it proportional to the defect. Since the defect can never exceed 180°, *there is an upper bound to the area of a triangle.*

Although Lobachevsky actually showed, by formal methods, that his geometry was consistent, this fact seems to have gone unrecognized at the time. The consistency of Bolyai-Lobachevsky geometry was not in fact publicly affirmed until Cayley, *Eugenio Beltrami* (1835–1900), *Felix Klein* (1849–1925) and others constructed *models* for it, that is, interpretations under which all its postulates could be seen to be true. In Klein's model we take a fixed circle C in the Euclidean plane and interpret *point* as *point in C*, *line* as *line in C*, and, glancing at the figure below, *distance between*

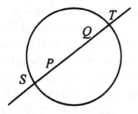

P and *Q* as a certain function of P and Q—actually the logarithm of the *cross-ratio* (*QPST*)—which becomes arbitrarily large as P approaches S or Q approaches T. This latter fact ensures that the "lines" in the model can be indefinitely "extended". In this model the parallel postulate obviously fails—there being many "lines" passing through a given point "parallel" to a given "line" in the sense that they fail to intersect *within C*. On the other hand the distance function can be chosen in such a way as to ensure that the other postulates of Euclidean geometry remain true.

The Klein model shows that the geometry of Bolyai and Lobachevsky is as consistent as that of Euclid. But which of the two is to be preferred as a description of the geometry of the *real world*? Notice that we can never determine by experiment whether there are one, or many, lines through a given point parallel to a given line. However, in Euclidean geometry the sum of the angles of a triangle is always 180°, while in Bolyai-Lobachevsky geometry it is always *less than* 180°. In the nineteenth century Gauss actually performed an experiment to determine which of these alternatives held. But although the result was, within the limits of experimental error, 180°, nothing was settled since, for small triangles (i.e. of terrestrial dimensions), the deviation from 180° might be so small as to be experimentally undetectable. That is, Bolyai-Lobachevsky and Euclidean geometry, although differing *in the large*, may coincide so closely *in the small* as to be empirically equivalent. So far as *local* properties of space are concerned, then, the choice between the two geometries can be made solely on the basis of simplicity and convenience.

The revolutionary importance of the discovery of noneuclidean geometry lay in the fact that it toppled Euclid's system as the immutable mathematical framework into which our knowledge of objective reality must be fitted.

While the *mathematician* may regard a "geometry" as being defined by any consistent set of postulates about "points", "lines", etc., the *physicist* will only find the result useful when the postulates in question conform with the behaviour of entities in the real world. Consider, for example, the statement *light travels in a straight line*. If this is regarded as the *physical definition* of a straight line, then the postulates of geometry must be chosen so as to correspond with the actual behaviour of light rays. The French mathematician *Henri Poincaré* (1854–1912) imagined a world confined within the interior of a circle *C*, in which the velocity of light at each point is inversely proportional to the distance of the point from the circumference of *C* (for example, *C* could be made of glass of suitably varying refractive index). It can then be shown that light rays will assume the form of *circular arcs* perpendicular at their extremities to the circumference of *C*, and thus that *Bolyai-Lobachevsky* geometry will prevail. Nevertheless, we can also arrange for *Euclidean* geometry to apply in this world: instead of regarding light rays as "noneuclidean" straight lines, we simply take them to be Euclidean circles normal to *C*. Thus we see that different geometries can describe the same physical situation, provided that physical entities (in the case just considered, light rays) are correlated with different notions in the geometries concerned.

In both Bolyai-Lobachevsky and Euclidean geometry it is tacitly assumed that *lines can be indefinitely extended*. But after Bolyai and Lobachevsky had revealed the possibility of constructing new geometries, it became natural to ask whether noneuclidean geometries could be constructed in which "straight lines" are not infinite, but *finite* and *closed*. Such geometries were first considered in 1851 by *Georg Friedrich Bernhard Riemann* (1826–1866). It turns out that geometries with closed finite lines can be constructed in a completely consistent way: we take as our "space" the surface S of a sphere in which we define *straight line* to mean *great circle on S*. Since every pair of great circles intersect (in two points), in this model there are *no* "parallel lines" at all.

Riemann extended the idea of his geometry by considering a "space" consisting of an arbitrary curved surface, and defining a "straight line" between two points on the surface to be the curve of shortest length or *geodesic* on the surface joining the points. In this case the deviation of the geodesics from Euclidean straightness provides a measure of the *curvature* of the surface. Riemann also extended this idea to three (and higher) dimensions, considering a geometry of (real) space analogous to the geometry of a surface, in which the curvature may change the character of the geometry from point to point. This is discussed further in the next chapter.

CHAPTER 8

THE DEVELOPMENT OF GEOMETRY, II

PROJECTIVE GEOMETRY

THE LAWS OF PERSPECTIVE DRAWING—the technique used to portray three dimensions on a two dimensional surface—have been studied by artists since the Stone Age. For example, a fifteen thousand year old etching of a herd of reindeer on a bone fragment discovered by archaeologists creates the impression of distance by displaying the legs and antlers as if seen beyond the fully sketched animals of the foreground. The main perspective problem encountered by *Egyptian* artists was the portrayal of a single important object with the necessary dimension of depth: this was achieved in an ingenious manner by drawing a combination of horizontal and side view. Thus, for instance, in drawing a Pharaoh carrying a circular tray of sacrificial offerings, the top view of the tray is shown in half display by means of a semicircle, and on this half-tray is presented the sacrificial food as it would appear from above. This stylized method of expressing a third dimension persisted in Egyptian drawing for three thousand years. In *America,* an arresting method of achieving this effect was created by northwest Indian artists who, in their drawings of persons or animals, present views of both front and left- and right hand sides. The figures are drawn as if split down the back and flattened like a hide, with the result that each side of the head and body becomes a profile facing the other. Landscapes drawn by *Chinese* artists create the impression of space and distance by skillful arrangement of land, water and foliage. In drawing buildings, however, it was necessary to display the parallel horizontal lines of the construction, and for this the technique of *isometric drawing* was used. This is a simulation of perspective drawing in which parallel lines are drawn parallel, instead of converging as in true perspective.

In *Europe,* it was not until the first half of the fifteenth century that Italian painters, through their introduction of the *horizon line* and *vanishing point,* transformed perspective drawing into an exact science. The fundamental principles were worked out by the Florentine architect *Filippo Brunelleschi* (1377–1446) and developed by the painter *Paolo Uccello* (1397–1475). The formulation of the laws of perspective quickly transformed painting in Renaissance Italy, and the technique of perspective became an essential constituent in the works of later Italian masters. In his great wall-painting *The*

Last Supper[1], for example, *Leonardo da Vinci* (1542–1519) employs perspective in a subtle way to draw the viewer's eye to the composition's centre.

The origins of *projective geometry* lie in the study of perspective. A painter's picture, or a photograph, can be regarded as a *projection* of the depicted scene onto the canvas or photographic film, with the painter's eye or the focal point of the camera's lens acting as the centre of projection. For example, suppose we take a photograph of a straight railway track, with equally spaced ties, going directly away from us. In the photograph the parallel lines of the rails appear to converge, meeting at a vanishing point or "point at infinity"; the equal spaces between the ties appear as unequal; and the right angles between the rails and the ties appear as acute. A circular pond in the landscape would appear as an ellipse. Nevertheless the geometric structure of the original landscape can still be discerned in the photograph. This is possible only if the original scene and its image have certain geometric properties in common, properties which, unlike lengths and angles, are preserved under the passage from the one to the other. The object of projective geometry is to identify and investigate these properties in a general setting.

We may describe the transformation from scene to image in the following way. Given any plane geometric figure (confining ourselves to plane figures for the sake of simplicity), let O be any point in the plane of the figure and imagine straight lines drawn from O to every point of the figure. Now allow this bundle of straight lines to be cut by any plane not passing through O. The resulting plane section of the lines through O yields a new figure in the cutting plane called the *projection* of the original figure. (In the case of our photograph, the original scene is the first figure, the photographic film is the cutting plane, and the focal point of the camera lens is the point O.) With each point or straight line of the original figure this process associates a definite point or straight line in the new figure; if a point P in the original figure lies on a straight line ℓ, in the new figure the associated point P' lies on the associated line ℓ'. Any mapping of one figure onto another by a projection of the kind we have just described, or the result of a finite sequence of such projections, is called a *projective transformation*. A projective transformation carries points into points and straight lines into straight lines in such a way as to preserve the incidence of points and lines. It does *not* preserve lengths or angles.

Two figures are *homologous* or *projectively equivalent* if each can be obtained from the other by a projective transformation. It is a fundamental fact of projective geometry that all the conic sections are projectively equivalent.

The set of all projective transformations furnishes another example of a *group*, since the result of successively performing any two of the operations in the set is still a transformation in the set; the set contains an identity transformation ("projection" onto the original plane); and each member of the set has an inverse whose composite with it in either order is the identity transformation (obtained by "reversing" the given projection). Projective geometry may be characterized completely as the study of those

---

[1] Happily, this painting, in a sadly deteriorated state for as long as anyone can remember, has recently (1999) been restored.

properties of figures which remain unchanged—are *invariant*—when acted upon by the elements of the group of projective transformations. Similarly, the set of all *rigid motions* in space forms a group: ordinary Euclidean geometry may then be identified as the geometry in which are studied those properties of figures, e.g., lengths and angles, which are invariant under the group of all such motions. As another example, the set of *conformal* (i.e., angle-preserving) *transformations* forms a group; *conformal geometry* —which arose in connection with cartography, a science rich in geometric possibilities—is the study of properties of figures which are invariant under all the transformations in this group. These are all special instances of a general principle— enunciated by Klein in his *Erlangen Program* of 1872—which asserts that corresponding to every group of transformations in space there is a "geometry" comprising those properties of figures which are invariant under all the transformations in the given group. This principle establishes a systematic link between geometry and algebra, and enables geometries to be classified by the properties of their corresponding groups. Thus, for example, since the group of rigid motions may be identified as a subgroup of the group of conformal transformations, which is in turn a subgroup of the group of projective transformations, ordinary Euclidean geometry is implicitly contained in conformal geometry, which is in turn contained in projective geometry. Each of the first two geometries may accordingly be obtained by specialization from the more general geometry containing it.

The first mathematician to take up the geometric problems suggested by perspective drawing was the Frenchman *Girard Desargues* (1591–1661) who was chiefly concerned to develop new methods for establishing properties of conics. In his major work on the subject, published in 1636, Desargues introduces, for each set of parallel lines, a ideal point—a *point at infinity* corresponding to the vanishing point of perspective drawing—which all the lines in the set have in common. He also introduces an ideal line—a *line at infinity* corresponding to the horizon line—and makes the assumption that all the new ideal points lie on that line. Thus in Desargues' geometry each "ordinary" line contains exactly one "ideal" point—its "point at infinity"; each pair of lines meets at, or determines, a unique point; and each pair of points determines a unique line.

Desargues' most famous result is his *Triangle Theorem* (see figure on following page).The theorem states: *If two triangles ABC and A'B'C' are situated in the same plane so that the straight lines joining corresponding vertices are concurrent in a point O, then the corresponding sides, if extended, will intersect in three collinear points.* The two triangles *ABC, A'B'C'* are said to be *in perspective* from O. This theorem clearly belongs to projective geometry, since the whole figure can be projected onto any other plane without affecting any of the features mentioned in the theorem's formulation. Its proof is based on this fact. For the points *Q* and *R* can then be "projected to infinity", causing the pairs of lines *AB, A'B'* and *AC, A'C'* to meet at ideal points, so rendering each pair parallel. It is not difficult to show that, as a result, *BC* and *B'C'* become parallel, so that their point of intersection *P* is now also an ideal point. Thus under projection the points *P, Q,* and *R* all lie on the line at infinity, and are, accordingly, collinear. Projecting back to the original plane, *P, Q,* and

R return to their original positions and remain collinear. So they must have been collinear in the first place. This proves the theorem.

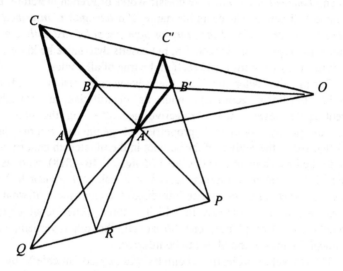

As noted by Desargues, the theorem remains true—and is, surprisingly perhaps, more easily proved—in three dimensions, when the triangles lie in different, nonparallel planes.

Desargues also established the fundamental fact that *cross-ratio* (a concept originally introduced by *Pappus of Alexandria c.*300 B.C.) *is invariant under projection.* The cross-ratio of four points A, B, C, D on a line ℓ is defined to be the quantity

$$(ABCD) \; = \; \frac{CA/CB}{DA/DB}$$

where a fixed direction on ℓ is taken as positive. Referring to the figure below, Desargues proves that $(ABCD) = (A'B'C'D')$. Pappus had established this fact only when A and A' coincide.

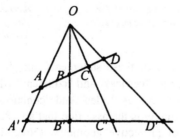

The next important contributor to projective geometry, *Blaise Pascal* (1623–1662) was at the same time a major literary figure and religious philosopher: his *Pensées* and *Lettres Provinciales* are classic works of French literature. Pascal is known for the famous theorem that bears his name: *if a hexagon is inscribed in a conic, the three points of intersection of the pairs of opposite sides are collinear.* Pascal did not supply an explicit proof of his theorem, but asserts that, since he knows it to be true for a circle, it must, by projection and section, be true of all conics.

During the latter half of the seventeenth century and throughout the eighteenth, coordinate geometry, with its quantitative and "analytic" methods, underwent rapid development. Projective geometry, on the other hand, with its emphasis on purely "synthetic" properties of geometric figures, languished. The stimulus that led to the revival of the subject in the nineteenth century was due in large measure to the French geometer *Gaspard Monge* (1746–1818), who, as an enthusiastic supporter of the French revolution, served as a technical advisor to Napoleon. In his *Traité de Géometrie* of 1799 Monge introduces the chief ideas of what was to become known as *descriptive geometry*, in which a three dimensional object is projected orthogonally onto two planes, one horizontal, the other vertical, so as to enable mathematical features of the object to be inferred.

The revival of projective geometry was carried out chiefly by Monge's pupils *Charles-Julien Brianchon* (1765–1864), *Lazare M. N. Carnot* (1753–1823), a major figure in the French revolution, and *Jean-Victor Poncelet* (1788–1867). Brianchon is remembered for the important theorem bearing his name: *the three diagonals joining opposite vertices of a hexagon circumscribed about a conic are concurrent.* Poncelet's work of 1822, *Traité des Propriétés Projectives des Figures*, is really the first systematic work on projective geometry, in which the subject is treated as an independent discipline with methods and goals of its own. It was Poncelet who established projective geometry in its modern sense as the study of properties of figures which are invariant under projection.

Poncelet was also the first to recognize what later became known—through the work of *Joseph-Diez Gergonne* (1771–1859) and *Jacob Steiner* (1796–1863)—as the *Principle of Duality*. This is best illustrated by writing the statements of Pascal's and Brianchon's theorems side by side:

Pascal's Theorem	*Brianchon's Theorem*
If the *vertices* of a hexagon *lie alternately on* two *straight lines*, the *points where opposite sides meet* are *collinear*.	If the *sides* of a hexagon *pass alternately through* two *points*, the *lines joining opposite vertices* are *concurrent*.

Here we observe that each of these statements may be obtained from the other by interchanging terms such as "vertices" and "collinear" which make reference to *points,* with corresponding terms such as "sides" and "concurrent" which make reference to *lines.* The Principle of Duality is the assertion that any theorem of projective geometry is likewise a theorem of projective geometry. (Note, incidentally, that the dual of Desargues' triangle theorem is precisely its converse!) The idea of duality, first discerned in projective geometry, has since been extended to other branches of mathematics, notably algebra.

The advances made in projective geometry in the first part of the nineteenth century also excited the interest of geometers whose work was primarily in the algebraic spirit. From their efforts issued the subject of *algebraic projective geometry*. Here a leading idea is that of a *homogeneous coordinate system*. The first such system was introduced by the German mathematician *August Ferdinand Möbius* (1790–1868) in his work *Der Barycentrische Calcul* of 1827. Möbius's idea was to start with a fixed triangle and to take as coordinates of any point P in the plane the amounts of mass which must be placed at each of the three vertices so as to make P the centre of mass or *barycentre* of the three masses (points lying outside the triangle being assigned some negative coordinates). Since multiplication of all three masses by the same nonzero constant does not change their barycentre, the coordinates are not unique, only their ratios being uniquely determined. Thus, as t varies over all nonzero real numbers, the triples (ta, tb, tc) all represent the same point, which justifies the use of the term "homogeneous" for the coordinate system.

The system of homogeneous coordinates normally employed by geometers today was invented by the German mathematician *Julius Plücker* (1801–1868). In his *Analytisch-geometrische Entwickelungen* of 1831, he takes the coordinates of any point P in the plane to be the signed distances (a, b, c) of P from the sides of a fixed triangle of reference. Since, again, the coordinates of a point can be multiplied by any nonzero constant without changing the point, Plücker went on to formulate what has become the standard definition of homogeneous coordinates in the plane, namely *ordered triples* (a, b, c) of real numbers not all zero, where the triples (ta, tb, tc) are taken to represent the same point as t varies over the nonzero real numbers. This is equivalent to replacing the usual rectangular coordinates (X, Y) by $(x/t, y/t)$, so that the equations of curves become *homogeneous* in x, y, and t, that is, all terms have the same degree. In Plücker's homogeneous coordinate system, ordinary points in the plane are represented by triples (a, b, c) with $c \neq 0$, ideal points lying on ordinary lines by triples $(a, b, 0)$ with $a \neq 0$, and points (necessarily ideal) lying on the line at infinity by triples $(0, b, 0)$ with $b \neq 0$. The *projective plane*, one of the most important concepts in projective geometry, is defined to be the set of points determined in this way.

DIFFERENTIAL GEOMETRY.

The introduction of the methods of the *calculus* into coordinate geometry during the seventeenth century issued in a branch of mathematics now known as *differential geometry*[2]. In contrast with Euclidean or projective geometry, which is concerned with the whole of a diagram or a figure, that is to say, with its *global* properties, differential

---

[2] The term "differential geometry" was introduced in 1894 by the Italian mathematician *Luigi Bianchi* (1856–1928).

geometry concentrates on its *local* properties, that is, those arising in the immediate neighbourhoods of points of the figure, and which may vary from point to point. Thus, for example, one of the central concepts of differential geometry is the *curvature* of a (plane) curve at a point, which is a quantity measuring the "sharpness of bending" of the curve in the neighbourhood of the point. More suggestively, if we think of the curve as being traced out by a moving particle, then the curvature at a given point is (proportional to) the rate at which the object's path is changing direction at that point. The curvature of a straight line at any point on it is then zero, as is, more generally, that at a point of inflection of any curve, while that of a circle is constantly equal to the reciprocal of the length of its radius. This latter fact suggests that we define the *radius of curvature* of a curve ℓ at a point *P* to be the *reciprocal* of the curvature at *P*. The radius of curvature is then the radius of the so-called *osculating,* or "kissing", circle—a

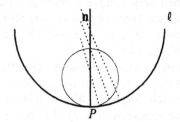

term first introduced by Leibniz—to the curve at *P*. This circle is the one which approximates best to the curve at *P;* its centre is called the *centre of curvature* of ℓ at *P*.

Many of the basic concepts of differential geometry were introduced by *Christiaan Huygens* (1629–1675) in his *Horologium Oscillatorium* of 1673, a work largely devoted, as its title suggests, to the theoretical design of accurate pendulum clocks. Huygens obtained his results by means of purely geometrical methods and it was in fact Newton who, in his *Geometria Analytica* of 1736, first employs the methods of the calculus in this area, obtaining results essentially identical to those of Huygens. Thus, for example, both Huygens and Newton show that the centre of curvature of a curve at a point *P* on it is the limiting position of the point of intersection of a fixed normal[3] **n** to a curve at *P* with an adjacent normal moving toward **n**, as in the figure above. They both show that the radius of curvature of the curve at the point (*x, y*) is, using the notation of the calculus,

$$\frac{[1 + (dy/dx)^2]^{3/2}}{d^2y/dx^2} \quad .$$

The first steps in *three-dimensional* differential geometry were taken by Clairault in his *Recherche sur les Courbes à Double Courbure* of 1731, a work treating of both curves and surfaces. Clairault termed space curves "curves of double curvature" because he recognized that the projection of a space curve onto each

---

[3] A *normal* to a curve at a point *P* on it is a straight line passing through *P* perpendicular to the tangent to the curve at *P*.

of two perpendicular planes would give rise to a pair of curves each possessing an independent curvature. He also saw that a space curve can have an infinity of normals in a plane perpendicular to the tangent at a point.

The next important advances in the theory of space curves were made in 1774 and 1775 by *Leonhard Euler* (1707–1783). He introduces the concept of *osculating plane* to a space curve at a point on it, that is, the plane lying closest to the curve in the neighbourhood of the point. Both the tangent and the osculating circle to the curve at the point lie in this plane. Euler defines the *curvature* of the curve at the point to be the reciprocal of the length of the radius of the osculating circle to the curve at the point.

In 1806 *Michel-Ange Lancret* (1774–1807), a student of Monge, introduced the scheme, now known as the *moving trihedron,* by means of which space curves are analyzed today. Lancret singled out three principal directions at each point P of a space

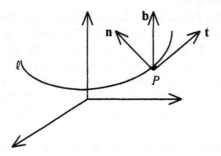

curve ℓ: the tangent **t**, the normal **n** in the osculating plane, and the common perpendicular to these, the *binormal* **b**. It is helpful to think of these as constituting a rigid frame of rectangular axes, with P as origin, moving forward and rotating as P traverses the curve with unit speed. The angular velocity of the frame about the binormal is then the curvature at P. The angular velocity of the frame about the tangent —the *torsion* of the curve at P—represents the rate at which the curve deviates from a plane at the point. The significance of the curvature and torsion of a curve is that together they provide a complete description of the curve: if we are given both at each point of the curve, then the curve is uniquely determined except for spatial position. This fact became clear when the theory of space curves was brought into essentially its modern form by *Augustin-Louis Cauchy* (1789–1857) in his *Leçons sur les Applications du Calcul Infinitésimal à la Géométrie* of 1826.

The Theory of Surfaces

The origins of the differential geometry of surfaces began in the seventeenth century with the study of *geodesics,* that is, curves of least length on a surface (initially, in accordance with the needs of navigation, that of the earth). In 1697 *Jean Bernoulli* (1667–1748) posed the problem of finding the shortest curve joining two points on a convex surface, and in 1698 observed that the osculating plane at any point of such a curve on a surface is normal to the curve at that point. Also in 1698 *Jacques Bernoulli*

(1655–1705) determined the geodesics on cylinders, cones, and surfaces of revolution. The general form of the equations for geodesics on surfaces was obtained by Euler in 1728.

The theory of surfaces was placed on a sound basis by Euler in his *Recherches sur la Courbure des Surfaces* of 1760. In this work he introduces the *principal curvatures* of a surface at a point. These are defined as follows. Given a surface S and a point P on it, the assemblage of planes through a normal n to S at P cuts S in a family of plane curves, each of which has a curvature at P. The two planes containing the curves with the largest and smallest curvatures can then be shown to be perpendicular to each other (assuming they are distinct). These two curvatures are then the principal curvatures of the surface S at P. It was shown in 1776 by *Jean-Baptiste Meunier* (1754–1793) that the only surfaces for which the two principal curvatures everywhere coincide are planes or spheres.

Euler was the first to study *developable* surfaces, that is, surfaces which can be flattened out onto a plane without distortion. This study had its origins in cartography, whose practitioners had come to recognize the awkward fact that a sphere cannot be cut and so flattened, thus making the task of mapping the earth's surface a subtle one. In his work *De Solidus quorum Superficium in Planum Explicare Licet* of 1771, Euler formulates necessary and sufficient conditions for a surface to be developable, and proves the striking result that the family of tangents to a space curve constitutes a developable surface.

In 1827 Gauss published his definitive paper on surfaces, *Disquisitiones Generales circa Superficies Curvas*, in which the study of the curvature of surfaces is raised to a new level. Gauss introduces a new measure of curvature, the *total* or *Gaussian curvature*, of a surface. To define Gaussian curvature, we consider a small region R on a surface S, and at each point of R we erect a normal to S. These normals

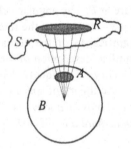

will form a cone, or solid angle, whose size is measured by the area of the region A in which a sphere B of unit radius intersects it. This size will depend both on the area of R and, crucially, on the extent to which it is curved. Thus the curvature of R may be characterized as the ratio of the size of the solid angle to the area of R. The total or Gaussian curvature of S at P is then defined to be the limit of this ratio as R shrinks to the point P.

Gauss proved that the total curvature of a surface at a point is the product of Euler's principal curvatures at that point. More significantly still, he established the following remarkable property of total curvature. Suppose that the surface has been stamped out from some flexible but inextensible material, a thin sheet of tin, for example, so that it can be bent into various shapes without stretching or tearing it. During this process the principal curvatures will change but, as was shown by Gauss, their product, and hence the total curvature, will *remain unchanged at every point.* This shows that two surfaces with different Gaussian curvatures are *intrinsically distinct,* the distinction consisting in the fact that the surfaces can never be deformed without stretching or tearing in such a way as to enable them to be superposed on one another. Thus, for example, a segment of the surface of a sphere can never be distorted in this manner so as to lie flat on a plane or on a sphere of different radius. Total curvature accordingly furnishes a measure of the curvature of a surface which is *intrinsic,* that is, not dependent on the fact that the surface is part of three-dimensional space. The idea that surfaces have intrinsic geometries was to serve as the basis for Riemann's far-reaching generalization of geometry, to which we now turn.

Riemann's Conception of Geometry

In his famous lecture of 1854, published as a paper in 1868, "On the Hypotheses which Lie at the Foundations of Geometry," Riemann introduces the idea of an intrinsic geometry for an arbitrary "space" which he terms a *multiply extended manifold.* Riemann conceives of a manifold as being the domain over which varies what he terms a *multiply extended magnitude.* Such a magnitude M is called *n-fold extended,* and the associated manifold *n-dimensional,* if n quantities—called *coordinates*—need to be specified in order to fix the value of M. For example, the position of a rigid body is a 6-fold extended magnitude because three quantities are required to specify its location and another three to specify its orientation in space. Similarly, the fact that pure musical tones are determined by giving intensity and pitch show these to be 2-fold extended magnitudes. In both of these cases the associated manifold is *continuous* in so far as each magnitude is capable of varying continuously with no "gaps". By contrast, Riemann terms *discrete* a manifold whose associated magnitude jumps discontinuously from one value to another, such as, for example, the number of leaves on the branches of a tree. Of discrete manifolds Riemann remarks:

> Concepts whose modes of determination form a discrete manifold are so numerous, that for things arbitrarily given there can always be found a concept...under which they are comprehended, and mathematicians have been able therefore in the doctrine of discrete quantities to set out without scruple from the postulate that given things are to be considered as being all of one kind. On the other hand there are in common life only such infrequent occasions to form concepts whose modes of determination form a continuous manifold, that the positions of objects of sense, and the colours, are probably the only simple notions whose modes of determination form a continuous manifold. More frequent occasion for the birth and development of these notions is first found in higher mathematics.

The size of parts of discrete manifolds can be compared, says Riemann, by straightforward counting, and the matter ends there. In the case of continuous manifolds, on the other hand, such comparisons must be made by *measurement.*

Measurement, however, involves superposition, and consequently requires the positing of some magnitude—not a pure number—independent of its place in the manifold.. Moreover, in a continuous manifold, as we pass from one element to another in a necessarily continuous manner, the series of intermediate terms passed through itself forms a one-dimensional manifold. If this whole manifold is now induced to pass over into another, each of its elements passes through a one-dimensional manifold, so generating a two-dimensional manifold. Iterating this procedure yields n-dimensional manifolds for an arbitrary integer n. Inversely, a manifold of n dimensions can be analyzed into one of one dimension and one of $n - 1$ dimensions. Repeating this process finally resolves the position of an element into n magnitudes. These ideas are not dissimilar to those put forward by Grassmann in his *Ausdehnungslehre.*

Riemann thinks of a continuous manifold as a generalization of the three-dimensional space of experience, and refers to the coordinates of the associated continuous magnitudes as *points*. He was convinced that our acquaintance with physical space arises only *locally*, that is, through the experience of phenomena arising in our immediate neighbourhood. Thus it was natural for him to look to differential geometry to provide a suitable language in which to develop his conceptions. In particular, the *distance* between two points in a manifold is defined in the first instance only between points which are at *infinitesimal* distance from one another. This distance is calculated according to a natural generalization of the distance formula in Euclidean space. In n-dimensional Euclidean space, we recall that the distance ε between two points P and Q with coordinates $(x_1,...,x_n)$ and $(x_1 + \varepsilon_1,..., x_n + \varepsilon_n)$ is given by

$$\varepsilon^2 = \varepsilon_1{}^2 + ... + \varepsilon_n{}^2. \tag{1}$$

In an n-dimensional manifold, the distance between the points P and Q—assuming that the quantities ε_i are infinitesimally small—is given by Riemann as the following generalization of (1):

$$\varepsilon^2 = \sum g_{ij}\varepsilon_i\varepsilon_j,$$

where the g_{ij} are functions of the coordinates $x_1,..., x_n$, $g_{ij} = g_{ji}$ and the sum on the right side, taken over all i, j such that $1 \leq i, j \leq n$, is always positive. The array of functions g_{ij} is called the *metric* of the manifold. In allowing the g_{ij} to be functions of the coordinates Riemann allows for the possibility that the nature of the manifold or "space" may vary from point to point, just as the curvature of a surface may so vary.

Riemann also extends Gauss's concept of total curvature of a surface to manifolds. In doing so his goal was to characterize Euclidean space, and, more generally, spaces which are *homogeneous* in that within them figures can be moved about without change of size or shape. Like total curvature of a surface, Riemann's notion of curvature of a manifold is an intrinsic property of the manifold (or, more precisely, of its metric); it is not required to think of the manifold as being situated in some manifold of higher dimension. Riemann introduces manifolds of *constant curvature*: by definition, in these spaces all measures of curvature are equal and remain unchanged from point to point, so yielding the required homogeneity property. In his paper Riemann states, but does not prove, that in a manifold of constant curvature the metric is given by

$$\varepsilon \;\; = \;\; \frac{\sqrt{(\Sigma \varepsilon_i{}^2)}}{(1 + \tfrac{1}{4}\alpha \, \Sigma x_i{}^2)}$$

Riemann observes that when α is positive we obtain a spherical space, when $\alpha = 0$, a Euclidean (flat) space, and when α is negative, a surface resembling the inside of a torus.

In the final section of his paper, Riemann applies his ideas to the problem of determining the structure of physical space. He points out that, as regards physical space, *infinitude* must be carefully distinguished from *boundlessness*. For example, the surface of a sphere is finite but unbounded and, for all we know, the same may be true of physical space. In any case it is the boundlessness, rather than the infinitude of space that is required for maneuvering in the external world, so that the former has a far greater empirical certainty than the latter. Moreover, if space has a constant positive measure of curvature, however small, then it would take the form of a spherical surface and would accordingly be finite and boundless.

Riemann concludes his discussion with the following words, the last sentence of which proved to be prophetic:

> While in a discrete manifold the principle of metric relations is implicit in the notion of this manifold, it must come from somewhere else in the case of a continuous manifold. Either then the actual things forming the groundwork of a space must constitute a discrete manifold, or else the basis of metric relations must be sought for outside that actuality, in colligating forces that operate on it. A decision on these questions can only be found by starting from the structure of phenomena that has been confirmed in experience hitherto...and by modifying the structure gradually under the compulsion of facts which it cannot explain...This path leads out into the domain of another science, into the realm of physics.

Riemann is saying, in other words, that if physical space is a continuous manifold, then its geometry cannot be derived *a priori*—as claimed, famously, by Kant—but can only be determined by experience. In particular, and again in opposition to Kant, who held that the axioms of Euclidean geometry were necessarily and exactly true of our conception of space, these axioms may have no more than approximate truth.

Riemann's final sentence proved to be prophetic because in 1916 his geometry—*Riemannian geometry*—was to provide the basis for a landmark development in physics, Einstein's celebrated *General Theory of Relativity*. In Einstein's theory, the geometry of space is determined by the gravitational influence of the matter contained in it, thus perfectly realizing Riemann's contention that this geometry must come from "somewhere else", to wit, from physics.

TOPOLOGY

Combinatorial Topology

The set of all *continuous* (*reversible*) *transformations* in space, i.e., deformations of figures or objects which occur without tearing anything apart, also forms a *group*, and the corresponding "geometry" is called *topology* (from Greek *topos,* "place"). By what we have already said, topology must then involve the consideration of properties of figures which are invariant under any continuous transformation. Calling two figures *topologically equivalent* if each one can be continuously and reversibly deformed into the other, we say that a property of a figure is *topological* if, when possessed by a given figure, it is also possessed by all figures topologically equivalent to the given one. *Topology* may now be broadly defined as the study of topological properties.[4] We see immediately that, in general, projective properties such as, e.g., being a triangle or a straight line are not topological, since a triangle is evidently topologically equivalent to any simple closed curve and a straight line to any open curve. (To see this, imagine both triangle and line made from cooked pasta or modelling clay.) Topological properties are grosser than projective ones, since they must stand up under arbitrary continuous deformations. So, for example, a topological property of the triangle is not its triangularity but the property of dividing the plane into two—"inside" or "outside"— regions, as well as the property that, if two points are removed, it falls into two pieces, while if only a single point is removed, one piece remains. As another example we may consider the properties of *one-sidedness* or *two-sidedness* of a surface. The standard one-sided surface—the so-called *Möbius strip* (or band), discovered independently in 1858 by Möbius and *J. B. Listing*[5] (1806–1882)—may be constructed by gluing together the two ends of a strip of paper after giving one of the ends a half twist. Both one- and two-sidedness are topological properties.

The historical record shows that the first theorem of an indisputably topological nature is that proved by Euler[6] in 1752. This is the assertion that, for any simple polyhedron (i.e., closed, convex, and without holes), the numbers v of vertices, e of edges and f of faces are related by the equation

$$v - e + f = 2. \tag{1}$$

---

[4]In view of the fact that a doughnut and a coffee cup (with a handle) are topologically equivalent, John Kelley famously defined a *topologist* to be someone who cannot tell the two apart. On this light-hearted note, the published phrase containing the maximum number of references to topology and geometry must surely be

> *On the analytic and algebraic topology of locally Euclidean metrization of infinitely differentiable Riemannian manifolds,*

which Tom Lehrer rattles off in his wittily irreverent "mathematical" song *Lobachevsky.*

[5] It was also Listing who, in his work *Vorstudien zur Topologie* of 1848, first uses the term "topology"; the subject being known prior to this as *analysis situs,* "positional analysis".

[6]The content of the theorem appears already to have been known to Descartes a century earlier.

Here is a proof of this assertion, based on that given by Cauchy in 1811. Imagine the given simple polyhedron to be hollow and made of thin rubber. Now cut out one of its faces and deform the remaining surface so as to stretch it out flat on a plane. In this way the network of vertices and edges of the original polyhedron is flattened out into a similar network on the plane, without changing the number of either. However, the number of polygons in the plane is one less than in the original polyhedron, since one face has been removed. We show that, for the plane network, $v - e + f = 1$, so that, when the excised face is counted, $v - e + f = 2$ for the original polyhedron.

First, we convert the plane network into a number of linked triangles by drawing all the diagonals in polygons which are not already triangles. This does not affect the value of $v - e + f$, because each time a diagonal is drawn, the values of e and f are both increased by 1, while that of v is unaffected. Next, from any triangle lying on the boundary of the network we remove that portion which does not belong to any other triangle. Any such triangle has either one or two edges on the boundary. In the first case, just the outer edge and the face of the triangle are removed, so that e and f are both decreased by 1, while v is unchanged, and accordingly $v - e + f$ remains the same. In the second case, the face, the two outer edges and one vertex of the triangle are removed, so that v and f are both diminished by 1, and e by 2, again leaving $v - e + f$ unaffected. Continuing in this way we can remove triangles on the boundary (which of course changes each time a triangle is removed) until finally only a single triangle remains, with its three edges, three vertices, and one face. For this simple network, $v - e + f$ clearly has value 1. Since the procedure of erasing triangles has not altered the value of $v - e + f$, this quantity must have value 1 for the original plane network, and hence also for the polyhedron with one face missing. So $v - e + f = 2$ for the original polyhedron, as claimed.

The *topological* nature of Euler's relationship (although Euler himself failed to recognize it) may be discerned through the fact that equation (1) will continue to hold when the polyhedron is subjected to an arbitrary continuous deformation. Under such a deformation, the edges will, in general, cease to be straight, and the faces cease to be flat, but, nevertheless, they will remain edges and faces. Thus the deformation will preserve both the number of edges and the number of faces, as well as, of course, the number of vertices, so that the relationship above will remain valid, even though the procedure will cause the surface of the polyhedron to become curved . The vertices, edges and faces of the original polyhedron may be considered to constitute a *map* drawn on the resulting (curved) surface. When all the faces of the map so constituted are *triangles*—curved or rectilinear—it is known as a *triangulation* of the surface. As we saw in the proof of Euler's theorem, the number $v - e + f$ for an arbitrary map is the same as for the triangulation obtained from it by drawing all the diagonals in its faces. Accordingly, as far as the value of $v - e + f$ is concerned, we lose nothing by confining our attention to triangulations. This fact is the basis of the *combinatorial method* in topology, in which the properties of the surface are investigated by means of one of its triangulations[7]. Of course, for this purpose one only considers properties of the triangulation which are *independent* of its particular identity and thus, being common to all triangulations of the given surface, express some property of the surface itself.

Euler's relationship (1) in fact yields one such property. Let us call the expression $v - e + f$, where v, e, and f are, respectively, the number of vertices, edges and triangles of the given triangulation its *Euler characteristic*. Then Euler's theorem can be taken as asserting that for all triangulations of a surface topologically equivalent to a sphere the Euler characteristic is *two*. As it turns out, all triangulations of any given surface have the same Euler characteristic, which we call the *Euler characteristic of the given surface*. This number is a *topological invariant* in the sense that its value is the same for all topologically equivalent surfaces. For example, the Euler characteristic of a cylinder or a torus (doughnut) is zero, and that of a figure eight shaped pretzel is -2. In general, the Euler characteristic of a surface with k "holes" is $2 - 2k$.

Another famous problem whose topological nature came later to be recognized is the *Königsberg Bridge Problem*. Königsberg lies on the banks of the river Pregel, which contains two islands linked to each other and to the banks in the following way:

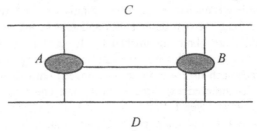

For many years the residents of Königsberg had tried without success to find a way of crossing all seven bridges exactly once on a continuous walk. Euler got wind of the problem and, in 1735, solved it in the negative. He achieved this by simplifying the formulation of the problem, replacing the land by points A, B, C, D and the linking bridges by arcs or lines as in the figure:

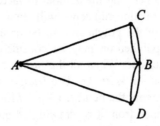

---

[7]This is an instance of the reduction of the *continuous* (in this case, a surface) to the *discrete* (in this case, the finite configuration provided by a triangulation). See Chapter 10.

The entire configuration is called a *graph*; its points are called *vertices,* and its lines and arcs *edges.* The problem of crossing the bridges is thus reduced to that of traversing the graph in one continuous sweep of the pencil without lifting it from the paper. Calling a vertex *odd* or *even* according as the number of edges leading from it is odd or even, Euler discovered that a graph can be continuously traversed in the manner specified if all its vertices are even. If the graph contains at least one, but no more than two odd vertices, it can be traversed in one journey, but it is not possible to return to the starting point. In general, if the graph contains $2n$ odd vertices, it will require exactly n journeys to traverse it. In the case of the Königsberg bridges $n = 2$.

It was Möbius who gave topology its first explicit formulation. In his *Theory of Elementary Relationships* of 1863 he proposed studying the relationship between two figures whose points can be placed in biunique correspondence in such a way that neighbouring points correspond to neighbouring points, that is, continuously. It is in this work that the technique of triangulation is first employed in a systematic way and used to show that any polyhedron may be systematically reduced to an assemblage of triangles. Möbius also showed that certain curved surfaces could be dissected and displayed as polygons with sides properly identified. Thus, for example, a double torus can be represented as the polygon below, where edges marked alike are to be identified:

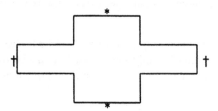

Another example of a topological property derives from the so-called *four-colour theorem*—first suggested by Möbius in 1840—which states that any map in the above sense drawn on the surface of a sphere (or a plane) can be coloured with at most four colours, in such a way that any two regions with a common boundary line are assigned different colours. The simplest map in the plane requiring four colours looks like this:

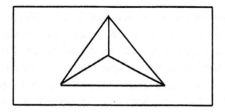

The problem of proving that four colours suffice for any map has an entertaining, if comparatively brief history, and includes what is probably the most infamous fallacious "proof" in the whole of mathematics. This was announced in 1879 by *A. B. Kempe* and for more than a decade was believed to be sound until *P. J. Heawood* spotted the flaw in 1890. By revising Kempe's proof Heawood was, however, able to establish that *five* colours are sufficient. In 1976, *W. Haken and K. Appel* finally established that four are enough, but only by resorting to the use of a computer program to check several hundred basic maps. It is with a certain reluctance that mathematicians have come to accept this "proof", since it cannot be examined in the traditional line-by-line manner that has always been taken for granted. The question of the legitimacy of "computer proofs" in mathematics has sparked a lively debate in recent years.

A major stimulus to the development of topology resulted from Riemann's work in the theory of functions of a complex variable in the middle decades of the nineteenth century. He introduced the concept of *connectivity* of a surface, defining it in the following way:

> If upon a surface there can be drawn *n* closed curves which neither individually nor in combination completely bound a part of the surface, but with whose aid every other closed curve forms the complete boundary of a part of the surface, then the surface is said to be $(n + 1)$-fold connected.

Riemann observes that the connectivity of a surface can be reduced by cutting it, giving as an example the torus, which is triply connected but can be reduced to a singly, or *simply* connected surface by means of one latitudinal and one longitudinal cut:

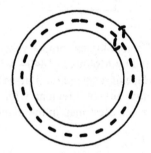

Riemann recognized that connectivity is a topological property and that it could be used to classify surfaces. In the latter half of the nineteenth century the term *genus* was introduced in connection with this classification. The genus of a surface is the largest number of nonintersecting simple closed curves that can be drawn on the surface without separating it: thus a sphere has genus 0, a torus genus 1 and, in general, a surface with *p* holes genus *p*. Riemann regarded it as intuitively evident that topologically equivalent surfaces have the same genus and observed that surfaces of genus 0 are topologically equivalent, each being mappable onto the surface of a sphere.

In 1882 Klein suggested a topological model for a surface of genus p which has become standard, namely, a sphere with p "handles" attached. It is clear that a closed surface of genus p can be continuously deformed into such an object, and so is topologically equivalent to it. Using this model it is quite easy to show that *the Euler characteristic of a surface S of genus p is* $2 - 2p$. For S may be taken to be a sphere with p handles, and each handle is attached to the sphere at two bounding curves. Cut the sphere along one of these bounding curves for each handle, and straighten the handles out. Each handle will now have a free edge bounded by a new curve having the same number of edges and vertices. Thus the additional edges and vertices exactly counterbalance one another, and so, since no new faces have been created, the Euler characteristic remains unchanged. Now flatten out the protruding handles, so that the resulting surface is simply a sphere from which $2p$ regions have been removed. Since the Euler characteristic of a sphere is 2, that of a sphere with $2p$ holes is clearly $2 - 2p$, hence also for the sphere with p handles, and any surface of genus p.

The problem of the topological equivalence of closed surfaces was investigated intensively throughout the latter half of the nineteenth century. In 1874 a major advance was made by Klein, who showed that two orientable closed surfaces of the same genus are topologically equivalent. Here by an orientable surface is meant one equipped with a triangulation whose constituent triangles can be oriented in such a way that any side common to two triangles has opposite orientations induced on it. Thus any two-sided surface such as a sphere or a plane is orientable, but a one-sided surface such as a Möbius strip is not.

By the end of the nineteenth century the only area of topology to have been more or less worked out was the topology of closed surfaces. Beginning in 1895 Poincaré began to develop the topological theory of spaces more general than surfaces, thereby establishing combinatorial topology as an autonomous branch of mathematics. In this general theory, later elaborated by the Dutch mathematician *L. E. J. Brouwer* (1882–1966), triangles are replaced by higher-dimensional objects known as *simplexes*—a one-dimensional simplex is a line segment, a two-dimensional simplex a triangle, a three-dimensional simplex a tetrahedron, and an n-dimensional simplex a generalized tetrahedron with $n + 1$ vertices. The role of triangulations is assumed by *complexes,* where a complex is defined to be an assemblage K of simplexes any two of which meet, if at all, in a common face, and every face of a simplex in K is also in K. Using these ideas the concept of connectivity and the Euler characteristic can be extended to spaces of higher dimensions.

Another important idea introduced by Poincaré is the so-called *fundamental group* of a space, the initial concept of what was to evolve into the subject known today as *algebraic topology.* The idea arises from considering the distinction between simply and multiply connected plane regions. In the interior of a circle, for instance, every closed curve can be contracted to a point, while in an annulus or circular ring a closed curve enclosing the inner circular boundary cannot be so contracted. Now consider an arbitrary region R in the plane—which we take as a typical "space"— and the closed curves that begin and end at a given point x_0 of R. Curves such as c and d in the figure below which can be continuously deformed into one another while remaining entirely within R are called *homotopic* and are considered to form one class. Roughly speaking,

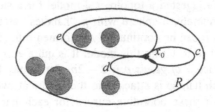

a class in this sense—a *homotopy* class—comprises those curves going round the same holes the same number of times. Thus the closed curves starting and ending at x_0 and not enclosing any holes form one homotopy class, while those such as e enclosing the two upper holes will form another. Two curves can be "merged" into the curve obtained by traversing first the one and then the other. This gives rise to an operation of *multiplication* on homotopy classes: the product of two homotopy classes C, D is to consist of all the curves obtained by merging a member of C with a member of D (in the order specified). It can now be shown that with this operation the collection of all homotopy classes forms a group, known as the *fundamental group* of the space R with base point x_0. Usually the structure of the fundamental group is independent of the choice of base point, and the group is simply called the *fundamental group* of the space in question. In this way an algebraic structure—a group—is associated with each space. The properties of the fundamental group reflect the connectivity of the space. Any simply connected space has a trivial fundamental group consisting of just the identity element, while the fundamental group of an annular region in the plane is isomorphic to the additive group of integers.

Yet another concept of algebraic topology which can be traced back to Poincaré is that of a *homology group*, a somewhat different device for charting the connectivity properties of the space. To explain the idea, consider the difference between the surface of a torus T and that of a sphere. It is intuitively obvious that any closed curve on the sphere cuts the surface into two separate regions. But this is not true in the case of the torus:

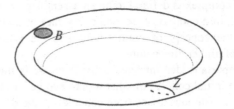

For cutting along the curve Z here does not disconnect the torus, which means that Z is not the boundary of a portion of T. But of course there are closed curves such as B which are boundaries.

Because the intuitive idea of a closed curve involves the notion that it "goes around something", a closed curve such as B or Z is called a *cycle* on T. While B and

Z are *simple* curves, i.e., do not cross themselves, we do not insist that all cycles be simple in this sense. The special cycles such as B—the *bounding* cycles—do not tell us very much about the structure of T, so for the moment we ignore them. It is the nonbounding cycles such as Z which are of interest.

Obviously there are infinitely many nonbounding cycles on the torus. But we shall see that these admit a simple description. First, the two cycles Z_1 and Z_2 in the figure below are not intrinsically different since they both go round the torus once latitudinally. More significantly, however, taken together they form the boundary of

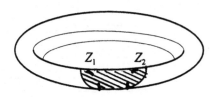

a portion of the torus, e.g. the shaded cylinder. The idea of taking cycles together suggests that they can be "added". So let us extend our definition of cycle to include all unions of finitely many closed curves. Thus the operation of union gives a well-defined sum of two cycles, and we see that the sum $Z_1 + Z_2$ of the two cycles considered above is a bounding cycle. This leads to the idea of calling two cycles Z and Z' *homologous* if $Z + Z'$ is a bounding cycle. This relation partitions cycles into mutually disjoint classes—the *homology classes*. The purely latitudinal cycles such as Z form one such homology class, the purely longitudinal ones another, and the bounding cycles still another. It is not difficult to show that every nonbounding cycle is homologous to a sum of a purely latitudinal cycle and a purely longitudinal one.

Now we can define an operation of *addition* on homology classes by stipulating that the sum of two such classes is to be the class consisting of the cycles obtained by adding a cycle of the first class to one of the second. With this operation the set of homology classes becomes a group—the *homology group* of the torus. It has 4 elements, namely the homology classes O, A, B, C consisting respectively of bounding cycles, purely latitudinal cycles, purely longitudinal cycles, and sums of cycles from A and B. It has neutral element O, and its addition table is:

$$A + A = B + B = C + C = O$$
$$A + B = B + A = C, \quad A + C = C + A = B, \quad B + C = C + B = A.$$

In a natural sense the two elements A and B may be said to *generate* the homology group of the torus; this corresponds to the fact that the torus has two "holes" around which curves can curves can go latitudinally or longitudinally. Topologists express this by saying that the Betti number—named after the Italian mathematician *Enrico Betti* (1823–1892)—of the torus is 2.

Using the notions of simplex and complex already mentioned, homology groups can be constructed for each space. The number of generators of this group—the Betti number of the space—is the number of "holes" in the space. Both homology and fundamental groups thus reflect the connectivity of the space. In general, homology groups are considerably easier to determine than fundamental groups. But the latter are more sensitive, as can be seen by considering the surface—depicted on the page following—obtained from a torus by removing an open disc D. While the fundamental group of this surface is clearly different from that of the torus, the homology group remains unchanged: the new "hole" is invisible as far as the latter is concerned.

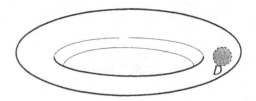

To conclude our account of combinatorial topology, we mention its most outstanding open problem, the so-called *Poincaré conjecture*. In 1904 Poincaré conjectured that every simply connected, closed orientable three-dimensional space is topologically equivalent to the three-dimensional sphere. This famous claim remains unresolved. Remarkably, however, the corresponding assertion for dimensions ≥ 4 has been established.

Point-set Topology

The key feature which distinguishes topology from other branches of geometry is its concern with *continuity*, a notion which in turn involves the idea of *proximity*. Thus a continuous function is, intuitively, one which does not pull "neighbouring" points of its domain "too far" apart. But what exactly do these terms mean? In the first decades of the twentieth century the use of *set theory* in developing *point-set* or *general topology* enabled these concepts to be rigorously defined, thereby placing the whole subject of topology on a solid basis.

The central concept of point-set topology is that of *topological space*. This concept may be defined in a variety of equivalent ways; the one given here—due originally to *Felix Hausdorff* (1868–1942)—is framed in terms of the concept of *neighbourhood*.

Suppose we are given a set X of elements which may be such entities as points in n-dimensional space, plane or space curves, complex numbers or quaternions, although we make no assumptions as to their exact nature. The members of X will be called *points*. Now suppose in addition that corresponding to each point x of X we are given a collection N_x of subsets of X, which we call the *neighbourhoods* of x, subject to the following conditions:

(i)   each member of N_x contains x;
(ii) the intersection of any pair of members of N_x includes a member of N_x;
(iii) for any U in N_x and any y in U, there is V in N_y such that $V \subseteq U$.

We may think of each U in N_x as determining a notion of proximity to x: the points of U are accordingly said to be *U-close* to x. Using this terminology, (i) may be construed as saying that, for any neighbourhood U of x, x is U-close to itself; (ii) that, for any neighbourhoods U and V of x, there is a neighbourhood W of x such that any point W-close to x is both U-close and V-close to x; and (iii) that, for any neighbourhood V of x, any point which is U-close to x itself has a neighbourhood all of whose points are U-close to x.

A *topological space* may now be defined to be a set X together with an assignment, to each point x of X, of a collection N_x of subsets of X satisfying conditions (i) – (iii). As examples of topological spaces we have: the real line **R** with the neighbourhood system at a point consisting of all open intervals with rational radii centred on that point; the Euclidean plane with the neighbourhood system at a point consisting of all open discs with rational radii centred on that point; and, in general, n-dimensional Euclidean space with the neighbourhood system at a point consisting of all open n-spheres centred on that point. Any subset X of any of these spaces becomes a topological space by taking as neighbourhoods the intersections with X of the neighbourhoods in the containing space.

Most topological notions can be defined entirely in terms of neighbourhoods. Thus a *limit point* of a set of points in a topological space is one each of whose neighbourhoods contains points of the set: a limit point of a set is thus a point which, while not necessarily in the set, is nevertheless "arbitrarily close" to it (lying on its boundary, for instance). A set is *open* if it includes a neighbourhood of each of its points, *closed* if it contains all its limit points, and *connected* if no matter how it is split into two disjoint sets, at least one of these contains limit points of the other. For example, any interval on the real line, and any Euclidean space, is connected in this sense.

The notions of continuous function and topological equivalence can also be defined in terms of neighbourhoods. Thus a function $f: X \to Y$ between two topological spaces X and Y is said to be *continuous*, if, for any point x in X, and any neighbourhood V of the image $f(x)$ of x, a neighbourhood U of x can be found whose image under f is included in V, in other words, such that the images of any point U-close to x is V-close to $f(x)$. In the case when X and Y are both the real line **R**, this condition translates into the famous ε, δ condition for continuity: for any x and for any $\varepsilon > 0$, there is $\delta > 0$ such that $|f(x) - f(y)| < \varepsilon$ whenever $|x - y| < \delta$. A *topological equivalence* or *homeomorphism* between topological spaces is a biunique function which is continuous in both directions, that is, both it and its inverse are continuous.

Metric spaces constitute an important class of topological spaces. A *metric* on a set X is a function d which assigns to each pair (x, y) of elements of X a nonnegative real number $d(x, y)$ in such a way that:

$$d(x, y) = d(y, x), \quad d(x, y) + d(y, z) \geq d(x, z).$$

We think of $d(x, y)$ as the *distance* between x and y. The first of these conditions then expresses the symmetry of distance, and the second—the *triangle inequality*—corresponds to the familiar fact that the sum of the lengths of two sides of a triangle is never exceeded by that of the remaining side. A set equipped with a metric is called a *metric space*. Each metric space may be considered a topological space in which a typical neighbourhood of a point x is the subset of X consisting of all points y at distance $< r$, for positive r. All of the examples of topological spaces given above are metric spaces, but not every topological space is a metric space. Metric spaces are the topological generalizations of Euclidean spaces.

In Euclidean spaces one has a natural definition of *dimension*, namely, the number of coordinates required to identify each point of the space. In a general topological space there is no mention of coordinates and so this definition is not applicable. In 1912 Poincaré formulated the first definition of dimension for a topological space; this was later refined by Brouwer, *Paul Urysohn* (1898–1924) and *Karl Menger* (1902–1985). The definition in general use today is formulated in terms of the *boundary* of a subset of a topological space, which is defined to be the set of all points of the space which are limit points both of the subset and its complement. Now the topological dimension of a subset M of a topological space is defined inductively as follows: the empty set is assigned dimension -1; and M is said to be n-dimensional at a point P if n is the least number for which there are arbitrarily small neighbourhoods of P whose boundaries in M all have dimension $< n$. The set M has topological dimension n if its dimension at all of its points is $\leq n$ but is equal to n at one point at least.

Dimension thus defined is a topological property. In 1911 Brouwer proved the important result that n-dimensional Euclidean space has topological dimension n. This means that Euclidean spaces of different dimensions cannot be topologically equivalent, even though—as Cantor showed (see Ch. 11)—they have the same "number" of points. Another important use of dimension theory is in the topological characterization of a curve. Menger and Urysohn defined a curve as a one-dimensional closed connected set of points, so rendering the property of being a curve a topological property.

CHAPTER 9

THE CALCULUS AND MATHEMATICAL ANALYSIS

THE ORIGINS AND BASIC NOTIONS OF THE CALCULUS

IN THE SIXTEENTH CENTURY the central concern of *physics* was the investigation of *motion*. The motion of a body in a given trajectory, a straight line, for instance, is determined by the way in which the *distance* changes with *time*. Thus *Galileo Galilei* (1564–1642) made the important discovery that the distance covered by a falling body increases in direct proportion to the square of the time. In general, descriptions of motion express the distance s in terms of the time t: Galileo's law may then be expressed in the form

$$s = \tfrac{1}{2}gt^2,$$

where g is the acceleration due to gravity (9.81 m/s^2). Here both s and t are examples of *variable quantities* or *variables*, since they take (continuously) varying values. The values of s and t are related in such a way that to each value of t there corresponds a unique value of s: this state of affairs is expressed—as we have observed—by asserting that the *dependent variable s* is a *function* of the *independent variable t*.

    The mathematicians of the seventeenth century found that the problems arising in the analysis of motion were closely related to certain geometric problems which had long resisted solution, and that these problems could be broadly divided into two classes. In the first class was to be found, for example, the problem of determining the velocity at any instant of a given nonuniform motion, or, more generally, finding the rate of change of a continuously varying magnitude, and the associated geometric problem of constructing the tangent to a given curve at a point on it. Efforts to solve problems of this kind led to what is now known as the *differential calculus*. The second class of problems included that of calculating the area of a given curved figure (the problem of *quadrature*) and the associated kinematical problem of finding the distance traversed in a nonuniform motion, or, more generally, determining the total effect of the action of a continuously changing magnitude. These problems issued in what is now known as the *integral calculus*. In this way two fundamental problems came to be identified—that of *tangents* and that of *quadrature*. The most important discovery in the mathematics arising from the analysis of motion was made independently by

Newton and Leibniz, who showed that in a certain sense these two problems are *inverse* to one another. Thus the differential and integral calculus came to be fused into a single mathematical discipline, known ever since as the *calculus*.

Let us first consider the problem of constructing the tangent to a curve at a given point. If we are given a curve and *two* points P and Q on it, it is a simple matter to draw the line or *chord* passing through them. As Q approaches P, the chord PQ,

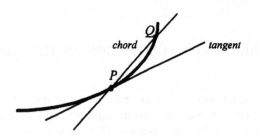

suitably produced, evidently provides a better and better approximation to the tangent at P. However, the chord PQ will differ (in general) from the tangent at P unless P and Q coincide. But when P and Q are coincident, they do not determine a unique line, since there are infinitely many straight lines passing through a given point. In the modern version of the differential calculus this problem is circumvented by obtaining the tangent to the curve at P as the line that *limits* the position of the chord PQ as Q approaches P—limiting in the sense that the angle between the tangent at P and the chord PQ can be made as small as we please by taking Q sufficiently close to P.

As an example, let us determine the slope of the tangent at the point $P = (x_0, y_0)$ on the parabola $y = x^2$. Let Q be the point $(x_0 + \delta x, y_0 + \delta y)$, where δx, δy are

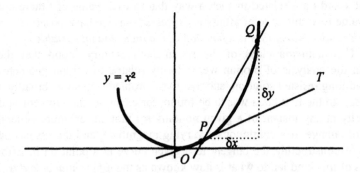

small increments in x_0, y_0 respectively. The slope of the chord PQ is given by the equation

$$\delta y / \delta x = \frac{(x_0 + \delta x)^2 - x_0^2}{\delta x} = 2x_0 + \delta x.$$

When Q is close to P, the increment δx is small, and by making it small enough we can bring Q as close to P as we please. Let PT be the line through P whose slope is $2x_0$. Then this line PT is the tangent at P, because the slope of PT is $2x_0$, that of PQ is $2x_0 + \delta x$, and we can take Q so close to P that the difference between these two slopes, namely δx, will be as small as we please. The slope of the tangent at (x_0, y_0) is, accordingly, that of PT, namely $2x_0$.

The process by which we passed from the slope $\delta y/\delta x$ of the chord PQ to the slope of the tangent at P is known as the *method of limits*. Thus, in our example above, where $\delta y/\delta x$ is equal to $2x_0 + \delta x$, the slope of the tangent, namely $2x_0$, is said to be the *limit* of the slope $2x_0 + \delta x$ as δx *tends to zero*. In this definition, there is present a quantity δx which we suppose assumes smaller and smaller values tending towards zero, that is, δx is a *variable tending to zero*. In a figurative sense, δx may be thought of as an *infinitesimal quantity*.

The concept of limit, one of the most important and far-reaching in mathematics, is extended to arbitrary functions as follows. Given a function $y = f(x)$, we say that A is the *limit of f(x)* as x *tends to a* if the difference between $f(x)$ and A remains as small as we please so long as x remains sufficiently near to a, *while remaining distinct from a*. The notation for a limit is *lim*, and the statement that the function $y = f(x)$ has limit A as x tends to a is then written

$$lim_{x \to a} f(x) = A. \tag{1}$$

In the nineteenth century, assertion (1) was provided with a watertight definition: *the function y = f(x) is said to have* limit A *as x tends to a if corresponding to any $\varepsilon > 0$ there is $\delta > 0$ for which*[1] $|f(x) - A| < \varepsilon$ *whenever* $0 < |x - a| < \delta$. Further, we say that $y = f(x)$ *has a limit as $x \to a$, or that the limit exists,* if there is a number A for which (1) holds.

Intuitively, the *gradient* of a function at a point is the slope of the tangent—assuming it to be well-defined—at that point to the curve determined by the function. This is given the following precise definition in terms of the limit concept. The function $y = f(x)$ is said to *have a gradient* or to be *differentiable* at $x = x_0$ if the limit as δx tends to 0 of the function

$$\frac{f(x_0 + \delta x) - f(x_0)}{\delta x}$$

exists. This limit, which is called the *gradient* of $y = f(x)$ at $x = x_0$, is often written in the less accurate but more suggestive form

$$lim_{\delta x \to 0} \delta y/\delta x. \tag{2}$$

---

[1] If r is a real number, we write $|r|$ for the *absolute value* of r, that is, $|r| = r$ if r is positive, and $|r| = -r$ if r is negative.

If f is differentiable at each point of its domain, then its gradient is a function of the argument x, being the function which associates, with each value of x, the slope of the tangent to the curve determined by $f(x)$. This function is called the *derivative* or *differential coefficient* of $f(x)$ with respect to x, and is written variously as $Df(x)$, Dy, $f'(x)$ or dy/dx, the latter notation being suggested by analogy with the expression in (2) above. The process of obtaining the differential coefficient is called *differentiation*.

In passing we note that a function differentiable at a point is also *continuous* at that point. Here $y = f(x)$ is said to be continuous at $x = a$ if

$$lim_{x \to a} f(x) = f(a).$$

This means simply that the curve determined by f does not have a "jump" or "gap" at $x = a$.

It is now straightforward to establish the usual laws for differentiation, for example:

$$D(ax^n) = nax^{n-1},$$

and, if $u(x)$ and $v(x)$ are both functions of x, then

$$D(uv) = uDv + vDu.$$

If we calculate the value of $\delta y/\delta x$ defined above for various functions $y = f(x)$, we find that in every case it consists of two terms, namely (i) the derivative, and (ii) a term of the form Ax_0. Accordingly, we have, for any value of x, an equation of the form

$$\delta y/\delta x = f'(x) + A x.$$

If we multiply this equation by x, we obtain

$$\delta y = f'(x)\delta x + A(\delta x)^2. \tag{3}$$

Now if δx is small, $(\delta x)^2$ is considerably smaller, and it follows that $f'(x)\delta x$ is a good approximation to δy. In recent years an approach to the calculus has been developed on the assumption of the presence of quantities δx which, while not actually being equal to zero, are nevertheless so small that their squares $(\delta x)^2$ are *literally* equal to zero. Quantities of this sort are called *square zero* or *nilsquare infinitesimals*: for any such quantity δx equation (3) becomes

$$f(x + \delta x) - f(x) = y = f'(x)\delta x.$$

In other words the change in the value of $f(x)$ attendant upon a nilsquare infinitesimal change δx in x is not merely approximately, but *exactly equal to $f'(x)\delta x$*. This enables the derivative $f'(x)$ to be *defined* as that number H which satisfies the equation

$$f(x + \delta x) - f(x) = H\delta x$$

for all nilsquare infinitesimal δx. This approach, which does not involve limits, considerably simplifies the development of the differential calculus: a further account of these ideas is given in Appendix 3.

The derivative of a function $y = f(x)$ of x is itself a function of x and so may itself have a derivative. For example, if $y = x^4$, then $Dy = 4x^3$, so that $D(Dy) = 12x^2$. The function $D(Dy)$, in this case, $12x^2$, is called the *second derivative* of y and is written D^2y, d^2y/dx^2, or $f''(x)$. Similarly, $D(12x^2)$, that is, $24x$, is called the *third derivative* of $y = x^4$. This process may be iterated indefinitely to yield, for each natural number n, the nth derivative D^ny or $f^{(n)}(y)$ of $y = f(x)$.

Use of the derivative enables various physical concepts to be given precise mathematical formulations. For example, suppose that at time t (seconds) from a given instant, a body of mass m (kilos), moving in a straight line, is at distance x (metres) from a fixed point O on the line. If the velocity of the body is v (metres per second) and the acceleration a (metres per second per second), then

$$v = dx/dt \quad a = dv/dt = d^2x/dt^2.$$

That is, velocity is the time derivative of distance, and acceleration is the time derivative of velocity, or the second time derivative of distance.

Newton called v the *fluxion* of x and a the *second fluxion* of x; x and v he also called the *fluents* (i.e., "flowing quantities"). His notation for a fluxion was a dot placed over (or alongside) the fluent; thus

$$v = \overset{\bullet}{x} \qquad a = \overset{\bullet}{v} = \overset{\bullet\bullet}{x} \ .$$

By his *second law of motion*, the *force F* in the direction of motion is the time rate of change of the *momentum mv*. Therefore

$$F = d(mv)/dt = mdv/dt = ma.$$

This equation is the mathematical presentation of Newton's second law.

An important application of the differential calculus is in determining *maximum and minimum values*. The method goes back in essence to Fermat. Suppose we are given a function $y = f(x)$; then the value of its derivative $f'(x_0)$ at x_0 is the slope of the (tangent to the) curve represented by $y = f(x)$ at x_0. If $f'(x_0) = 0$, then the curve is said to have a *stationary point* at x_0. It is obvious that a point at which f assumes a maximum or minimum value must be a stationary point: this fact makes the determination of these values a simple matter, as the following example illustrates.

Suppose that a closed cylindrical can is to be made to contain a given volume V, and we want to determine the shape of the can so that the amount of material used in its manufacture is a minimum. We proceed as follows. Writing r for the radius of the can's base and h for its height, its total surface area is then

$$S = 2rh + 2r^2.$$

Thus the problem is to minimize $2rh + 2r^2$, subject to the constraint $V = r^2h$. Here we may simplify the algebra by saying that the problem is to minimize $rh + r^2$ subject to $r^2h = c$. If we write

$$f(r) = r^2 + rh = r^2 + c/r,$$

then

$$f'(r) = 2r - c/r^2,$$

so that $f'(r) = 0$ if $2r^3 = c$. Thus a stationary point for $f(r)$ is obtained when $2r^3 = c$, i.e. when $2r^3 = r^2h$, so that $h = 2r$. This value is a minimum because it is plain that $f(r)$ increases indefinitely both when r tends to zero and when r increases indefinitely. We conclude that the surface area, and hence also the amount of material, is a minimum when *the can's height and base diameter coincide.*

We turn now to the problem of *quadrature*, that is, of determining the area enclosed in, or bounded by, a curve. Thus (to follow Newton) let CPD be the graph of the function $y = f(x)$, $OA = a$, $AC = f(a)$, $OM = x$, $MP = f(x)$. Let z denote the area $AMPC$; this area may be thought of as being generated by a line parallel to the y-axis

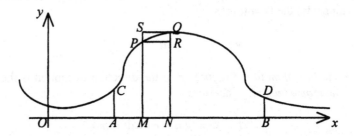

which sets out from the position AC and moves to the right. Thus z is a function of x which is zero when $x = a$. We wish to calculate z, and to do this we find dz/dx. (For simplicity we assume that $f(x)$ is always positive.) When x is increased by the amount δx (= MN), the area z increases by the amount δz (= area $MNQP$). Draw PR, SQ parallel to MN. Then $MNQP$ is equal to the rectangle $MNRP$, together with the area PRQ, which is less than the area of the rectangle $PRQS$; therefore

$$\delta z = MP. \ \delta x + \text{a quantity} < RQ. \ \delta x,$$

so that

$$\delta z/ \ \delta x = MP + \text{a quantity} < RQ.$$

As δx tends to zero, so does RQ, and accordingly

$$dz/dx = MP = f(x). \tag{4}$$

Note incidentally that the use of nilsquare infinitesimals makes this demonstration even simpler, for when δx is nilsquare, the area of PRQ, being proportional to $(\delta x)^2$, is *literally* zero. Therefore $z = MP$. x: since this holds for arbitrary x, equation (4) again follows (see Appendix 3).

The area z under the curve $y = f(x)$ is therefore *that function of x whose derivative is $f(x)$* (and which vanishes when $x = a$). We write

$$z = \int f(x)dx,$$

reading the right-hand side of this equation as the (indefinite) *integral of $f(x)$*; the operation of forming the integral is called *integration*. By definition the operation of differentiation is *inverse* to that of integration, i.e.,

$$d/dx[\int f(x)dx] = f(x).$$

The theorem that we have just demonstrated, namely, that the derivative of the function representing the area under a curve given by $y = f(x)$ is just $f(x)$ itself, is called the *Fundamental Theorem of the Calculus*: it forges the link between the two types of problem—that of tangents and that of quadrature—with which we began our discussion.

Returning to our figure on the previous page, there is a special notation for the integral that is equal to the area $ABDC$, namely,

$$\int_a^b f(x)dx.$$

This is called a *definite integral*; it is computed by finding the indefinite integral of $f(x)$, replacing x by b, then by a, and subtracting the second result from the first.

For example, the area under the curve $y = 4x^3 + 3x^2 + 2x$ between $x = 0$ and $x = 2$ is calculated as follows. Since

$$d/dx[x^4 + x^3 + x^2] = 4x^3 + 3x^2 + 2x,$$

it follows from the definition of the integral that

$$\int (4x^3 + 3x^2 + 2x)dx = x^4 + x^3 + x^2,$$

so the area in question is

$$\int_0^2 (4x^3 + 3x^2 + 2x)dx = (x^4 + x^3 + x^2)_{x=2} - (x^4 + x^3 + x^2)_{x=0}$$

$$= 2^4 + 2^3 + 2^2 - 0 = 28.$$

The interpretation of the area z under a curve will depend on the nature of the quantities represented by x and y. Here are some examples.

(i) If OM or x represents *time* and MP *velocity*, then the area z represents the *gain in distance* during the time represented by AM.

(ii) If OM represents time and MP *acceleration*, then z represents the *velocity* gained in the time represented by AM.

(iii) If OM represents the distance a *force* moves its point of application and MP the force, then z represents the *work done* by the force in moving its point of application through the distance represented by AM.

(iv) If OM represents time and MP the *temperature* in a room, then z represents the total *energy* expended in heating the room during the time represented [2] by AM.

There is an alternative way of arriving at the definite integral, namely as *the limit of a sum*, a method which, in essence at least, goes back to the ancient Greek method of exhaustion but which in its modern form is due to Riemann. Thus, consider once again the curve defined by the function $y = f(x)$ between $x = a$ and $x = b$. We will see first how the area under the curve, which we shall now denote by A, can be approximated by a *sum of rectangles*. We divide the interval from a to b into N small subintervals (x_m, x_{m+1}), of equal width $(b - a)/N$, with $x_0 = a$ and $x_n = b$, and define $\delta x = x_m - x_{m-1} = (b - a)/N$. We choose δx as the common length of the bases of

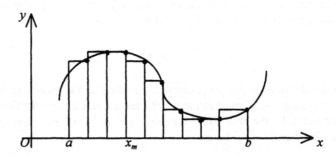

our "approximating rectangles". As their heights we choose the value of $y = f(x)$ at the right-hand endpoint of the subinterval. Thus the sum of the areas of these rectangles will be

$$S_N = f(x_1)\delta x + f(x_2)\delta x + ... + f(x_N)\delta x,$$

---

[2] Thus the energy expended in heating a room over a stretch of time depends only on the *average* temperature during that time: when one is out it is cheaper to turn the thermostat down and allow the room to cool than to maintain it at a constant temperature, even though the process of reheating it on one's return may require a disconcertingly steady expenditure of heat energy.

which we abbreviate as [3]

$$S_N = \sum_{j=1}^{N} f(x_j)\delta x.$$

Now we form a sequence of such approximations in which N increases indefinitely, so that the number of terms in each sum increases while, owing to the presence of the factor $\delta x = (b - a)/N$, each single term $f(x_j)\delta x$ tends to zero. By taking N sufficiently large, we can make S_N differ from the area A by an amount as small as we please. This we express briefly by saying that A is the *limit* of the S_N as N tends to infinity. We may then define the definite integral $\int_a^b f(x)dx = A$ as the *limit of the sums* $S_N = \sum_{j=1}^{N} f(x_j) x$ as N tends to infinity.

Leibniz symbolized this passage to the limit from the approximating sum S_N to A by replacing the summation sign $\sum$—which in his time was written "S"—by the symbol $\int$ (a stylized "S"), and the difference symbol δ by the symbol d. This notation is standard today.

The *natural logarithm* $\log(x)$ is an important example of a function defined by means of integration. It is defined for positive x by

$$\log(x) = \int_1^x (1/t)dt.$$

The number e—like π, one of the fundamental constants of mathematics—is defined to be that number whose logarithm is 1, that is,

$$1 = \int_1^e (1/t)dt;$$

e has the value 2.7182818.... The function e^x, also written $\exp(x)$, is called the *exponential function*; it can be shown that e^x is the function *inverse* to $\log(x)$, that is, $e^{\log(x)} = \log(e^x) = x$. The exponential function has the key property of being equal to its own derivative: $D(e^x) = e^x$.

The definition of the integral as the limit of a sum of plane areas can be extended to three (and higher) dimensions so as to enable the differential calculus to be employed in determining areas of curved surfaces and volumes of solids. This definition also provides the basis for the general theories of integration which have been developed in the present century.

---

[3] Here the symbol $\sum_{j=1}^{N}$ signifies the sum of all the expressions obtained by allowing j to assume successively the values 1, 2, 3,.., N.

MATHEMATICAL ANALYSIS

Within both the differential and integral calculus the limit concept plays a central role: the derivative of a function is the limit of a quotient, and the definite integral is the limit of a sum. A characteristic feature of the limit concept is that it involves the idea of an *infinite process*: in forming a limit as x tends to zero, say, we imagine x becoming smaller and smaller endlessly without ever actually vanishing. The use of infinite processes is the hallmark of the extensive area of mathematics known as *mathematical analysis*, or *analysis* for short. Analysis may be considered, along with geometry and algebra (here taken to include number theory), as one of the three main divisions of mathematics. Analysis is, perhaps, the largest of these divisions, and includes many topics besides the calculus; only a few of these will be touched on here. We defer until Chapter 11 our discussion of *set theory*, the branch of mathematics concerned with the infinite in its purest form, and which may be considered to have emerged from analysis.

Infinite Series

Let $a_0, a_1, a_2, \ldots$ be a sequence of real numbers. We say that the sequence—which we shall denote by $\langle a_n \rangle$—*converges to limit a as n tends to infinity*—written $a_n \to a$ as[4] $n \to \infty$—if the differences between a_n and a become arbitrarily small as n becomes arbitrarily large. To be precise, $\langle a_n \rangle$ converges to limit a if, corresponding to any $\varepsilon > 0$ there is an integer N such that $|a_n - a| < \varepsilon$ for all $n \geq N$. In the opposite event, the sequence is said to *diverge*.

Given a sequence $\langle a_n \rangle$, the sequence $\langle S_n \rangle$ defined by

$$S_n = \Sigma_{j=1}^{n} a_j = a_1 + a_2 + \ldots + a_n$$

is called the sequence of partial sums of the *infinite series*

$$\Sigma_{n=1}^{\infty} a_n,$$

sometimes written simply

$$\Sigma a_n.$$

If $S_n \to s$ as $n \to \infty$, the series is said to *converge* to the *sum s*, and we write

$$s = \Sigma_{n=1}^{\infty} a_n = a_1 + a_2 + \ldots + a_n + \ldots .$$

---

[4] The sign "∞" to denote an infinite number was introduced by Wallis in his work *De Sectionibus Conicis* of 1655. It has been conjectured that Wallis, who was a classical scholar, adopted this sign from the late Roman symbol "∞" (possibly a form of "M") for 1000.

In the opposite event the series is said to *diverge*.

The most familiar type of convergent series is the *geometric series*

$$\sum_{n=1}^{\infty} x^n$$

with fixed x satisfying $|x| < 1$. In this case the partial sums satisfy

$$S_n = \sum_{j=1}^{n} x^j = 1 + x + \ldots + x^n = \frac{1 - x^{n+1}}{1 - x}.$$

If $|x| < 1$, then $x^{n+1} \to 0$ as $n \to \infty$, so that

$$S_n \to 1/(1 - x) \quad \text{as } n \to \infty.$$

Accordingly $\sum_{n=1}^{\infty} x^n$ converges with sum $1/(1 - x)$ provided $|x| < 1$.

The ancient Greeks were conversant with the idea that an infinite series could converge, even if they lacked a precise definition of the concept. In particular they had grasped the fact that

$$\tfrac{1}{2} + \tfrac{1}{4} + \tfrac{1}{8} + \ldots + 1/2^n + \ldots = 1.$$

In Chapter 6 of his *Physics* Aristotle observes, in effect, that such a series should have a sum.

Infinite series make an occasional appearance in medieval mathematics. *Nicolas Oresme* (c. 1323–1382), for example, observed that the so-called *harmonic series*

$$\tfrac{1}{2} + \tfrac{1}{3} + \tfrac{1}{4} + \ldots$$

is *divergent*. For it can be replaced by the series of lesser terms

$$(\tfrac{1}{4} + \tfrac{1}{4}) + (\tfrac{1}{8} + \tfrac{1}{8} + \tfrac{1}{8} + \tfrac{1}{8}) + \ldots,$$

which yields as many terms of magnitude $\tfrac{1}{2}$ as one pleases.

In 1593 Viète derived a formula for the sum of an infinite geometric series, and in 1647 *Gregory of St. Vincent* (1584–1687), in his *Opus Geometricum*, is the first to state explicitly that a convergent infinite series represents a magnitude, which he calls the limit of the series.

The principal use of infinite series has been to represent *functions*, and thereby to provide a means for calculating their values. In the mid-seventeenth century, for instance, Newton obtained a series for the natural logarithm

$$\log(1 + x) = x - \tfrac{1}{2} x^2 + \tfrac{1}{3} x^3 - \ldots,$$

which converges for $|x| \leq 1$. This is a typical example of what is known as a *power series*. Newton also obtained the power series

$$e^x = 1 + x + x^2/2! + x^3/3! + ..., \tag{1}$$

so that in particular

$$e = 1 + 1 + 1/2! + 1/3! + ...,$$

as well as power series for the trigonometric functions[5]

$$\sin x = x - x^3/3! + x^5/5! - ... \tag{2}$$
$$\cos x = 1 - x^2/2! + x^4/4! - ... . \tag{3}$$

An important general method, still very much in use today, for representing a function as a power series was developed by *Brook Taylor* (1685–1731) in his *Methodus Incrementum Directa et Inversa* of 1715. Building on the work of Newton and James Gregory, Taylor obtains for the function $y = f(x)$ the equation

$$f(x + h) = f(x) + f'(x)h + f''(x)h^2/2! + f'''(x)h^3/3! + ...$$

Putting $x = 0$ yields the equation

$$f(h) = f(0) + f'(0)h + f''(0)h^2/2! + f'''(0)h^3/3! + ...$$

This is known as *Maclaurin's theorem,* after *Colin Maclaurin* (1698–1746). In obtaining their series neither Taylor nor Maclaurin employ rigorous argument, nor do they consider the question of convergence. Maclaurin derives his series by means of the method of *undetermined coefficients*. He begins by assuming that $f(x)$ can be expanded as a series

$$f(x) = A + Bx + Cx^2 + Dx^3 + ...$$

Then by successive term-by-term differentiation he obtains

$$f'(x) = B + 2Cx + 3Dx^2 + ...$$
$$f''(x) = 2C + 6Dx + ...$$

Putting $x = 0$ in each equation then determines $A, B, C, D, ...$.

Euler was the master of infinite series. These he would manipulate with great flair, only rarely paying attention to the question of convergence. In his *Introductio in*

---

[5] Here $\sin x$ and $\cos x$ are given as functions of a variable angle x measured in *radians*, where a radian is the the angle subtended at the centre of a circle by an arc equal to the radius. Thus π radians $= 180°$.

Analysin Infinitorum of 1748 are to be found many examples of his skill in this regard. One of his most famous results obtained by the formal manipulation of infinite series relates the exponential and logarithmic functions through complex numbers and is known as the *Euler identity*, namely,

$$e^{ix} = \cos x + i \sin x. \tag{4}$$

This memorable equation may be derived by noting that, if we substitute ix for x in the series expansion (1) for e^x and use the series expansions (2) and (3) for $\sin x$ and $\cos x$, we get

$$\begin{aligned} e^{ix} &= 1 + ix - x^2/2! - ix^3/3! + x^4/4! + ix^5/5! + \ldots \\ &= (1 - x^2/2! + x^4/4! - \ldots) + i(x - x^3/3! + x^5/5! - \ldots) \\ &= \cos x + i \sin x. \end{aligned}$$

If we substitute π for x in (4), and note that $\cos \pi = -1$, $\sin \pi = 0$, we obtain one of the most remarkable equations in the whole of mathematics, namely,

$$e^{i\pi} + 1 = 0.$$

Of course, equations such as this only become meaningful when quantities such as $e^{i\pi}$ are properly defined: this was not carried out until the nineteenth century.

Using his freewheeling methods, Euler was able to achieve results which baffled his contemporaries. Among these was the summing of the series $1/1^2 + 1/2^2 + 1/3^2 + \ldots$. To sum this, Euler began with the equation (2) for $\sin x$; the equation $\sin x = 0$ can then be thought of—after dividing through by x—as the infinite polynomial equation

$$0 = 1 - x^2/3! + x^4/5! - \ldots$$

or, replacing x^2 by y,

$$0 = 1 - y/3! + y^2/5! - \ldots .$$

From the theory of algebraic equations it is known (and not difficult to show) that if the constant term in a polynomial equation is 1, then the sum of the reciprocals of the roots is the negative of the coefficient of the linear term, which in this case is $1/3!$. Moreover, the roots of the equation $\sin x = 0$ are known to be $\pi, 2\pi, 3\pi, \ldots$; accordingly, the roots of the equation in y are $\pi^2, (2\pi)^2, (3\pi)^2, \ldots$. Therefore

$$1/6 = 1/\pi^2 + 1/(2\pi)^2 + 1/(3\pi)^2 + \ldots,$$

or

$$\pi^2/6 = 1/1^2 + 1/2^2 + 1/3^2 + \ldots . \tag{5}$$

Using the cosine series (4) in place of the sine series, Euler similarly obtained the result

$$\pi^2/8 = 1/1^2 + 1/3^2 + 1/5^2 + \dots . \tag{6}$$

Subtracting equation (5) from twice equation (6) gives

$$\pi^2/12 = 1/1^2 - 1/2^2 + 1/3^2 - 1/4^2 + \dots .$$

These arguments, while hardly rigorous, are nothing less than spectacular. In the nineteenth century, the great majority of Euler's results were derived by rigorous, but less entertaining methods.

An important class of infinite series are the so-called *trigonometric series*, which had first arisen in connection with the analysis of planetary orbits. By a trigonometric series is meant any series of the form

$$\tfrac{1}{2}a_0 + \Sigma_{n=1}^{\infty}(a_n\cos nx + b_n\sin nx).$$

Each of the sine and cosine functions in this series has a wave-like or "sinusoidal" graph of the form

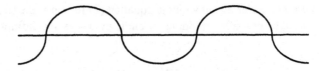

Thus the function represented by such a series is a "superposition" of such waves.

Trigonometric series were investigated in the eighteenth century by Euler, *Jean–le–Rond d'Alembert* (1717–1783), Clairaut, and Lagrange, but it was *Joseph Fourier* (1768–1831)—in his papers of 1807, 1811, and above all in his celebrated work *Théorie Analytique de la Chaleur* of 1822—who first advanced the claim that an *arbitrary* function $y = f(x)$ can be represented as a trigonometric series, thus introducing the branch of mathematics that has become known as *harmonic analysis*. Fourier obtains the expressions (in which he was anticipated by Euler)

$$a_n = 1/\pi\int_0^{2\pi} f(x)\cos nx\, dx, \quad b_n = 1/\pi\int_0^{2\pi} f(x)\sin nx\, dx$$

for the coefficients of the trigonometric series expansion of $f(x)$. Fourier's work was so influential that the trigonometric series representing a function is called its *Fourier expansion* and the coefficients of this series its *Fourier coefficients*.

Fourier did not supply a rigorous proof that the trigonometric series associated with a function converges to the value of the function, and later work showed that for certain functions this was not in fact the case. It was Dirichlet who, in 1828, gave the first rigorous proof of convergence of the Fourier expansion of a function subject to certain restrictions. He showed that, if $y = f(x)$ is continuous at each point of its domain, periodic of period 2π, that is, $f(x + 2\pi) = f(x)$, and $\int_{-\pi}^{\pi} f(x)dx$ is finite, then the Fourier expansion of f always converges to $f(x)$.

The mathematicians of the eighteenth century had been rather cavalier in their use of infinite series, and by the century's end some dubious or downright absurd results obtained by their means provoked enquiries as to the validity of operating with them. In the forefront of these investigations was Cauchy, who in his *Cours d'Analyse* of 1821 gave the rigorous definition of convergence of an infinite series employed today, and who also formulated the purely internal criterion for convergence which has come to bear his name (although it was known earlier to Bolzano). This states that a necessary and sufficient condition that an infinite series Σa_n converge is that, for any given value of p, the difference between the partial sums S_n and S_{n+p} tends to zero as n tends to infinity.

While the injection of rigor into the theory of infinite series administered in the nineteenth century may have tamed the subject, it has remained colourful. The resulting precision also made possible the discovery of new and striking facts about infinite series. *Riemann's theorem* is an example. This stems from the observation that certain series containing positive and negative terms exist with the property that, while they converge, the series obtained by making all their terms positive do not. For instance,

$$1 - \tfrac{1}{2} + \tfrac{1}{3} - \tfrac{1}{4} + \ldots$$

converges to log 2, but, as we have seen, the corresponding series of positive terms

$$1 + \tfrac{1}{2} + \tfrac{1}{3} + \tfrac{1}{4} + \ldots$$

diverges to infinity. A series of this type is called *conditionally convergent*. Riemann's theorem asserts that, given any conditionally convergent series, its terms can be rearranged so that the resulting series converges to *any value whatever*. Thus it can be shown that

$$1 + \tfrac{1}{3} - \tfrac{1}{2} + 1/5 + 1/7 - \tfrac{1}{4} + \ldots = 3/2 \log 2,$$

and

$$1 - \tfrac{1}{2} - \tfrac{1}{4} + \tfrac{1}{3} - 1/6 + \tfrac{1}{8} + \ldots = \tfrac{1}{2} \log 2.$$

Differential Equations

In the seventeenth and eighteenth centuries mathematicians applied the calculus to a broader and broader range of physical problems. The mathematical formulation of many of these problems involved what came to be known as *differential equations,* that is, equations involving derivatives of a function or functions, from which the functions themselves are to be determined.

A general problem of this sort is the following. Consider a material particle of mass m moving along a line, and let x denote its distance at time t from some given fixed point on the line. Let us assume that the motion is caused by some force F, the value of which depends on x, on the velocity dx/dt, and on the time t; to indicate this dependence we write $F = F(x, dx/dt, t)$. According to Newton's second law of motion, the action of the force F on the particle produces an acceleration $a = d^2x/dt^2$ with $F = ma$, and so we obtain the differential equation

$$m d^2x/dt^2 = F(x, dx/dt, t). \tag{1}$$

More specifically, let us suppose that the particle is moving in a resisting medium, for example in a liquid or a gas, under the influence of the elastic force of two springs, acting in accordance with *Hooke's law* (after *Robert Hooke,* 1635–1703), which states that the elastic force of a spring acts towards the equilibrium position and is proportional to the deviation therefrom. Assuming that the equilibrium position is

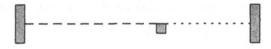

at $x = 0$, the elastic force is then proportional to $-bx$, where b is some positive constant. We also suppose that the resistance of the medium is proportional to the speed of the motion, that is, equal to $-adx/dt$, where a is some positive constant. In that case the force F acting on the particle is $-bx - adx/dt$ and so equation (1) assumes the form

$$m\, d^2x/dt^2 + a\, dx/dt + bx = 0.$$

This is the *differential equation of motion* of the particle. Its solution $x = x(t)$ gives the position x of the particle as a function of the time t.

Differential equations also arise in the determination of the *form of curves*. For example, in 1690 Jacques Bernoulli posed the problem of finding the curve assumed by a flexible inextensible cord hanging freely from two fixed points, the curve Leibniz named the *catenary.* This problem had been considered as far back as the fifteenth century by Leonardo da Vinci; Galileo mistakenly thought that the curve was a parabola. Huygens showed that this was not correct and that the curve would assume a parabolic form only when the combined weight of the cord and any objects suspended from it is uniform per horizontal run, as in a suspension bridge. In 1691 Leibniz,

Huygens and Jean Bernoulli published independent solutions to the problem, the latter employing the methods of the calculus. Writing $y = f(x)$ for the equation of the curve, Bernoulli obtains the differential equation

$$1 + (dy/dx)^2 = a^2 y,$$

and gives the correct solution

$$y = \tfrac{1}{2}a(e^{x/a} - e^{-x/a}).$$

Perhaps the most celebrated solution to a set of differential equations is Newton's derivation of Kepler's laws of planetary motion. Formulated in 1609 and 1619 by *Johannes Kepler* (1571–1630), these laws are the following three assertions:

1. *The planets move about the sun in elliptical orbits with the sun at one focus*[6].
2. *The radius vector joining a planet to the sun sweeps over equal areas in equal intervals of time.*
3. *The square of the time of one complete revolution of a planet about its orbit is proportional to the cube of the orbit's semimajor axis.*

In his *Philosophiae Naturalis Principia Mathematica*—universally known as the *Principia*—of 1687, Newton formulates his laws of motion and the inverse square law of gravitational attraction between two bodies, and from them deduces Kepler's laws, at the same time creating the science of *celestial mechanics*. Although Newton casts his work in the form of classical Greek geometry, it is almost certain that he obtained his results in the first instance by the use of the calculus, which means, in essence, that he solved the differential equations of motion of a body moving under a central force.

Partial differential equations are also frequently encountered in physics. These are equations involving *partial derivatives*. If $f(x, y, z, ...)$ is a function of the variables $x, y, z, ...,$ the partial derivative $\partial f/\partial x$ of f with respect to one of these variables, x, say, is obtained by fixing the values of the remaining variables $y, z, ...,$ and then differentiating the resulting function of x. Similarly one obtains higher partial derivatives $\partial^2 f/\partial x^2$ and mixed partial derivatives $\partial^2 f/\partial x \partial y$. It is a fundamental property of partial differentiation that the order in which the differentiations are made is irrelevant: thus, for example, $\partial^2 f/\partial x \partial y = \partial^2 f/\partial y \partial x$.

One of the first, and undoubtedly most important, partial differential equations—the *wave equation*—made its appearance in the eighteenth century in connection with the analysis of a vibrating string, a violin string, for example. In 1746 d'Alembert gave the correct form of the equation governing a flexible continuous vibrating string:

---

[6] An ellipse may be defined as the locus of all points the sum of whose distances from two given points is constant: these two points are the *foci* of the ellipse.

$$T \, \partial^2 u/\partial x^2 = \rho \, \partial^2 u/\partial t^2, \tag{2}$$

where $u = u(x, t)$ is the displacement of the string from its equilibrium position at time t and position x along it, T is the tension in the string and ρ is its density. D'Alembert shows that (2) has a solution of the form

$$u(x, t) = \varphi(ct + x) + \psi(ct - x),$$

where $c = \sqrt{(T/\rho)}$ and φ, ψ are arbitrary twice differentiable functions. This solution may be considered as the superposition of two "standing waves", each moving in opposite directions along the string.

In 1748 Euler gave a solution to the wave equation in the form of a trigonometric series

$$u(x, t) = \Sigma A_n \sin n\pi x/\ell \cos n\pi ct/\ell.$$

This solution presents the characteristic vibrations of the string in terms of a superposition of infinitely many elementary wave motions of sinusoidal form.

Later the wave equation was extended to three spatial dimensions, taking the form

$$\partial^2 u/\partial x^2 + \partial^2 u/\partial y^2 + \partial^2 u/\partial z^2 = c^2 \, \partial^2 u/\partial t^2.$$

This equation, one of the most important in mathematical physics, governs the vibrations of solid bodies and the propagation of waves such as sound and light.

Another important partial differential equation in mathematical physics is the *potential equation*, also known as *Laplace's equation*, which takes the form

$$\partial^2 V/\partial x^2 + \partial^2 V/\partial y^2 + \partial^2 V/\partial z^2 = 0.$$

This equation, which was given in 1789 by *Pierre Simon de Laplace* (1749–1827), governs the *gravitational potential* $V(x, y, z)$ outside a gravitating body. The potential function $V(x, y, z)$ is characterized by the property that its partial derivatives with respect to the three spatial coordinates x, y and z are the three components of the gravitational force exerted by the body at the point (x, y, z).

Other important partial differential equations arise in connection with *fluid flow*. In a paper of 1755 entitled *General Principles of the Motion of Fluids*, Euler derives his famous equations of flow for perfect nonviscous fluids. If we write u, v, w for the components of the velocity of the fluid at time t in the x, y and z directions, p for the pressure and ρ for the density of the fluid, the first of Euler's equations for compressible flow takes the form

$$\partial p/\partial x = -\rho[u\partial u/\partial x + v\partial u/\partial y + w\partial u/\partial z + \partial u/\partial t],$$

with corresponding expressions for $\partial p/\partial y$ and $\partial p/\partial z$. Euler also obtains a general form of the so-called *equation of continuity* which had been derived previously by d'Alembert for incompressible flow, and which expresses the fact that matter is neither created nor destroyed in fluid motion. Euler's version of the equation is:

$$\partial \rho/\partial t + \partial(\rho u)/\partial x + \partial(\rho v)/\partial y + \partial(\rho w)/\partial z = 0.$$

For incompressible flow ρ is constant and we obtain d'Alembert's form of the equation

$$\partial u/\partial x + \partial v/\partial y + \partial w/\partial z = 0.$$

If we follow Euler by introducing a function S, called the *velocity potential*, whose partial derivatives with respect to x, y, and z are u, v and w, respectively, we see that S satisfies Laplace's equation. This equation is also characteristic of fluid flow.

Finally, we must mention the *heat equation* which was obtained by Fourier in his *Théorie Analytique de la Chaleur*. By physical arguments, Fourier showed that the temperature T at a point (x, y, z) at time t in a homogeneous and isotropic body satisfies the equation

$$\partial^2 T/\partial x^2 + \partial^2 T/\partial y^2 + \partial^2 T/\partial z^2 = k^2\,\partial T/\partial t,$$

where k is a constant whose value depends on the material of the body. When T is independent of the time, $\partial T/\partial t = 0$ and the equation reduces to Laplace's equation. In one dimension Fourier gave a solution to the equation in the form of a trigonometric series with exponential coefficients

$$T = \Sigma b_n \exp(-n^2\pi^2/k^2\ell^2)t \sin n\pi x/\ell.$$

The presence of the exponential terms here reflects the fact that, as a body cools, its temperature falls off as e^{-kt} with the time t.

Complex Analysis

One of the most significant and original mathematical creations of the nineteenth century was the *theory of functions of a complex variable*, or *complex analysis* for short. Now, strictly speaking, a function of a complex variable is just a function $f: \mathbf{C} \to \mathbf{C}$, so that, for each $z = x + iy$, $f(z)$ is a complex number whose real and imaginary parts are uniquely determined by x and y. In other words, there are functions $u(x, y)$ and $v(x, y)$ of two real variables x and y such that $f(x + iy) = u(x, y) + iv(x, y)$. Thus, according to this definition, a complex function is essentially nothing more than a pair of real functions.

Both Cauchy and Riemann independently saw that this definition, while perfectly legitimate, was too broad to be of much practical value. They proposed instead that attention be confined to complex functions which are *differentiable* in the same sense as real functions. Thus a function $f: \mathbf{D} \to \mathbf{C}$, where $\mathbf{D}$ is a region in the complex plane is said to be *differentiable* at a point z_0 of $\mathbf{D}$ if

$$\frac{f(z) - f(z_0)}{z - z_0} \tag{1}$$

tends to a unique limit as $z \to z_0$. A function which is differentiable in this sense at every point of a region $\mathbf{D}$ is said to be *analytic* (also: *regular* or *holomorphic*) on $\mathbf{D}$.

Since analyticity requires that (1) has a unique limit when $z - z_0$ tends to zero not merely through purely real or purely imaginary values, but *along any path whatsoever*, it is evidently a very strong condition. Cauchy and Riemann independently discovered that the real and imaginary parts $u(x, y)$ and $v(x, y)$ of an analytic function satisfy what are now known as the *Cauchy–Riemann equations*, viz.,

$$\partial u/\partial x = \partial v/\partial y \qquad \partial u/\partial y = -\partial v/\partial x.$$

And conversely, if u and v are continuous and have continuous partial derivatives satisfying the Cauchy-Riemann equations throughout a region, then $f = u + iv$ is analytic in that region.

From the Cauchy-Riemann equations we deduce, by partial differentiation,

$$\partial^2 v/\partial x \partial y = \partial^2 u/\partial x^2 = -\partial^2 u/\partial y^2 \text{ and } \partial^2 u/\partial x \partial y = -\partial^2 v/\partial x^2 = \partial^2 v/\partial y^2.$$

Hence both u and v satisfy the two-dimensional version of Laplace's equation, that is

$$\partial^2 V/\partial x^2 + \partial^2 V/\partial y^2 = 0.$$

It follows that, by separating any analytic function into its real and imaginary parts, two solutions of Laplace's equation are immediately obtained. It is this fact that makes complex analysis so useful in the solution of problems in mathematical physics, especially in the theory of fluid flow.

Cauchy proved what is perhaps the single most important result of complex analysis, now known as the *Cauchy integral theorem*. Given a curve C in a region of the complex plane, the *integral* of an analytic function f along C may be defined in a manner analogous to that of the definite integral of a real function. Namely, if a, b are the endpoints of the curve, we subdivide it into n segments by means of the successive

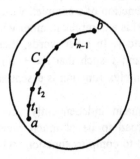

points $a = t_0, t_1, t_2, ..., t_n = b$ and form the sum

$$S_n = \sum_{i=1}^{n} f(t'_i)(t_i - t_{i-1}),$$

where t'_i denotes any point on C lying between t_{i-1} and t_i. If we now make the subdivision finer and finer by allowing the number of points to increase without limit in such a way that the greatest of the distances $|t_i - t_{i-1}|$ tends to zero, then S_n tends to a limit which is independent of the choice of the points t_i and t'_i. This limit is the integral of $f(z)$ along C and is written

$$\int_C f(z)dz.$$

Now the Cauchy integral theorem asserts that, if $f(z)$ is analytic in a simply connected region **D** of the complex plane, and C, C' are any two simple curves in **D** with the same end points, then

$$\int_C f(z)dz = \int_{C'} f(z)dz.$$

That is, *the integral is independent of the path*. In particular, if C is a closed curve, then

$$\int_C f(z)dz = 0.$$

Cauchy's theorem has many consequences, of which we mention in conclusion one of the most important. This is that every analytic function can be expanded as a *power series*. Precisely, if $f(z)$ is analytic on and inside a simple closed curve C, then, for any given point a inside C, there are complex numbers $b_1, b_2, ...$ for which

$$f(z) = f(a) + b_1(z - a) + b_2(z - a)^2 + ...$$

whenever z is inside C. Notice that, since the coefficients $b_1, b_2, ...$ depend only on a, a function analytic in a region can be represented as a single power series, not just locally, but throughout the entire region. This shows once again what a powerful condition analyticity is.

CHAPTER 10

THE CONTINUOUS AND THE DISCRETE

THE RELATIONSHIP BETWEEN THE IDEAS of *continuity* and *discreteness* has played no less important a role in the development of mathematics than it has in science and philosophy. Continuous entities are characterized by the fact that they can be *divided indefinitely* without altering their essential nature. So, for instance, the water in a bucket may be continually halved and yet remain water[1]. Discrete entities, on the other hand, typically cannot be divided without effecting a change in their nature: half a wheel is plainly no longer a wheel. Thus we have two contrasting properties: on the one hand, the property of being indivisible, separate or discrete, and, on the other, the property of being indefinitely divisible and continuous although not actually divided into parts.

Now one and the same object can, in a sense, possess both of these properties. For example, if the wheel is regarded simply as a piece of matter, it remains so on being divided in half. In other words, the wheel regarded as a wheel is discrete, but regarded as a piece of matter, it is continuous. From examples such as these we see that continuity and discreteness are complementary attributes originating through the mind's ability to perform acts of abstraction, the one arising by abstracting an object's divisibility and the other its self-identity.

In mathematics the concept of whole number provides an embodiment of the concept of pure discreteness, that is, of the idea of a collection of separate individual objects, all of whose properties—apart from their distinctness—have been refined away. The basic mathematical representation of the idea of continuity, on the other hand, is the geometric figure, and more particularly the straight line. Continuity and discreteness are united in the process of measurement, in which the continuous is expressed in terms of separate units, that is, numbers. But these separate units are unequal to the task of measuring in general, making necessary the introduction of fractional parts of the individual unit. In this way fractions issue from the interaction between the continuous and the discrete.

---

[1]For the purposes of argument we are ignoring the atomic nature of matter which has been established by modern physics.

A most striking example of this interaction—amounting, one might say, to a collision—is the Pythagorean discovery of incommensurable magnitudes. Here the realm of continuous geometric magnitudes resisted the Pythagorean attempt to reduce it to the discrete form of pure number. The "proportions" invented by Eudoxus to resolve the problem were in essence an extension of the idea of number—i.e., of the discrete— adequate to the task of expressing the relations between continuous magnitudes. Many centuries later, in the modern era, the concept of real number finally crystallized as a complete solution to the problem of representing continuous magnitudes as numbers.

The opposition between the continuous and the discrete also arises in physicists' account of the nature of matter, or substance. An early instance of this opposition—perhaps the first—in physics occurred in Greece in the third century B.C. with the emergence of two rival physical theories, each of which became the basis of a fully elaborated physical doctrine. One is the *atomic theory*, due to *Leucippus* (fl. 450 B.C.) and *Democritus* (b. *c*.470 B.C.) The other—the *continuum theory*—is the creation of the *Stoic* school of philosophy and is associated with the names of *Zeno of Cition* (fl. 250 B.C.) and *Chrysippus* (280–206 B.C.).

The continuum of Stoic philosophy is an infinitely divisible continuous substance which was conceived as providing the ultimate foundation for all natural phenomena. In particular the Stoics held that *space* is everywhere occupied by a continuous invisible substance which they called *pneuma* (Greek: "breath"). This pneuma—which was regarded as a kind of synthesis of air and fire, two of the four basic elements, the others being earth and water—was conceived as being an elastic medium through which impulses are transmitted by wave motion. All physical occurrences were viewed as being linked through tensile forces in the pneuma, and matter itself was held to derive its qualities form the "binding" properties of the pneuma it contains.

The atomists, on the other hand, asserted that material form results from the arrangement of the atoms—the ultimate building blocks—to be found in all matter, that the sole form of motion is the motion of individual atoms, and that physical change can only occur through the mutual impact of atoms.

A major problem encountered by the Stoic philosophers was that of the nature of *mixture*, and, in particular, the problem of explaining how the pneuma mixes with material substances so as to "bind them together". The atomists, with their granular conception of matter, did not encounter any difficulty here because they could regard the mixture of two substances as a combination of their constituent atoms into a kind of lattice or mosaic. But the Stoics, who regarded matter as continuous, had difficulty with the notion of mixture. For in order to mix fully two continuous substances, they would either have to interpenetrate in some mysterious way, or, failing that, they would each have to be subjected to an infinite division into infinitesimally small elements which would then have to be arranged, like finite atoms, into some kind of discrete pattern.

This controversy over the nature of mixture shows that the problem of continuity is intimately connected with the problems of infinite divisibility and of the infinitesimally small. The mixing of particles of finite size, no matter how small they may be, presents no difficulties. But this is no longer the case when we are dealing with a continuum, whose parts can be divided *ad infinitum*. Thus the Stoic philosophers were confronted with what was at bottom a *mathematical* problem.

In fact, the problem of infinite divisibility had already been posed in a dramatic but subtle way more than a century before the rise of the Stoic school, by *Zeno of Elea* (fl. 450 B.C.), a pupil of the philosopher *Parmenides* (fl. 500 B.C.), who taught that the universe was a static unchanging unity. Zeno's arguments take the form of *paradoxes* which are collectively designed to discredit the belief in motion, and so in any notion of change. We consider, in modern formulation, three of these paradoxes, which are, perhaps, the most famous illustrations of the opposition between the continuous and the discrete. The first two of these, both of which rest on the assumption that space and time are continuous, purport to show that under these conditions continuous motion engenders, *per impossibile,* an actual infinity.

The first paradox, that of the *Dichotomy*, goes as follows. Before a body in motion can reach a given point, it must first traverse half of the distance; before it can traverse half of the distance, it must traverse one quarter; and so on *ad infinitum*. So, for a body to pass from one point to another, it must traverse an infinite number of divisions. But an infinite number of divisions cannot be traversed in a finite time, and so the goal cannot be reached.

The second paradox, that of *Achilles and the Tortoise*, is the most famous. Achilles and a tortoise run a race, with the latter enjoying a head start. Zeno asserts that no matter how fleet of foot Achilles may be, he will never overtake the tortoise. For, while Achilles traverses the distance from his starting-point to that of the tortoise, the tortoise advances a certain distance, and while Achilles traverses this distance, the tortoise makes a further advance, and so on *ad infinitum*. Consequently Achilles will run *ad infinitum* without overtaking the tortoise.

This second paradox is formulated in terms of two bodies, but it has a variant involving, like the *Dichotomy,* just one. To reach a given point, a body in motion must first traverse half of the distance, then half of what remains, half of this latter, and so on *ad infinitum*, and again the goal can never be reached. This version of the *Achilles* exhibits a pleasing symmetry with the *Dichotomy*. For the former purports to show that a motion, once started, can never stop; the latter, that a motion, once stopped, can never have started.

The third paradox, that of the *Arrow*, rests on the assumption of the *discreteness of time*. Here we consider an arrow flying through the air. Since time has been assumed discrete we may "freeze" the arrow's motion at an indivisible instant of time. For it to move during this instant, time would have to pass, but this would mean that the instant contains still smaller units of time, contradicting the indivisibility of the instant. So at this instant of time the arrow is at rest; since the instant chosen was arbitrary, the arrow is at rest at any instant. In other words, it is always at rest, and so motion does not occur.

Let us examine these paradoxes more closely. In the case of the *Dichotomy*, we may simplify the presentation by assuming that the body is to traverse a unit spatial interval—a mile, say—in unit time—a minute, say. To accomplish this, the body must first traverse half the interval in half the time, before this one-quarter of the interval in one-quarter of the time, etc. In general, for every subinterval of length $1/2^n$ ($n = 1, 2,$

3,…), the body must first traverse half thereof, i.e. the subinterval of length $1/2^{n+1}$. In that case both the total distance traversed by the body and the time taken is given by the convergent series

$$\Sigma_{n=0}^{\infty}(1/2^{n} - 1/2^{n+1}) = \Sigma_{n=0}^{\infty} 1/2^{n+1} = 1,$$

as expected. So, *contra* Zeno. the infinite number of divisions is indeed traversed in a finite time.

More troubling, however, is the fact that these divisions, of lengths …, $1/2^{n}$, … , $1/2^{3}$, $1/2^{2}$, ½ constitute an infinite *regression* which, like the negative integers, *has no first term*. Zeno seems to be inviting us to draw the conclusion that it cannot be supplied with one, so that the motion could never get started. However, from a strictly mathematical standpoint, there is nothing to prevent us from placing 0 before all the members of this sequence, just as it could be placed, in principle at least, before all the negative integers. Then the sequence of correlations … , $(1/2^{n}, 1/2^{n})$, …, (½,½), (1, 1) between the time and the body's position is simply preceded by the correlation (0, 0), where the motion begins. There is no contradiction here.

In the case of the *Achilles*, let us suppose that the tortoise has a start of 1000 feet and that Achilles runs ten times as quickly. Then Achilles must traverse an infinite number of distances—1000 feet, 100 feet, 10 feet, etc.—and the tortoise likewise must traverse an infinite number of distances—100 feet, 10 feet, 1 foot, etc.—before they reach the same point simultaneously. The distance of this point in feet from the starting points of the two contestants is given, in the case of Achilles, by the convergent series

$$\Sigma_{n=0}^{\infty} 1000.10^{-n} = 1111\ 1/9,$$

and in the case of the tortoise

$$\Sigma_{n=0}^{\infty} 100.10^{-n} = 111\ 1/9.$$

And, assuming that Achilles runs 10 feet per second, the time taken for him to overtake the tortoise is given, in seconds, by the convergent series

$$\Sigma_{n=0}^{\infty} 100.10^{-n} = 111\ 1/9,$$

so that, again *contra* Zeno, Achilles overtakes the tortoise in a finite time.

Although the use of convergent series does confirm what we take to be the evident fact that Achilles will, in the end, overtake the tortoise, a nagging issue remains. For consider the fact that, at each moment of the race, the tortoise is somewhere, and, equally, Achilles is somewhere, and neither is ever twice in the same place. This means that there is a biunique correspondence between the positions occupied by the tortoise and those occupied by Achilles, so that these must have the same number. But when Achilles catches up with the tortoise, the positions occupied by the latter are only *part* of those occupied by Achilles. This would be a contradiction if one were to insist, as did Euclid in his *Elements,* that the whole invariably has more terms than any of its parts. In fact, it is precisely this principle which, in the nineteenth century, came to be repudiated for infinite sets[2] such as the ones encountered in Zeno's paradoxes. Once this principle is abandoned, no contradiction remains.

The paradox of the *Arrow* can be resolved by developing a theory of *velocity*, based on the differential calculus. By definition, (average) velocity is the ratio of distance travelled to time taken. It will be seen at once that in this definition *two* distinct points in space and *two* distinct points in time are required. Velocity at a point is then defined as the *limit* of the average velocity over smaller and smaller spatiotemporal intervals around the point. According to this definition, a body may have a nonzero "velocity" at each point, but at each instant of time will not "appear to be moving".

Although, as we have seen, Zeno's paradoxes can be resolved from a strictly *mathematical* standpoint, they present difficulties for understanding the nature of *actual* motion which have persisted to the present day.

The Stoic philosophers, as well as mathematicians such as Eudoxus, grasped fully the idea of passage to the limit or convergence of purely *spatial* quantities. But the notion of convergence of points or intervals of *time* eluded them, because getting at this notion involves the idea of a *functional correspondence* between time and space, a conception which never received adequate formulation in ancient Greek science. Nevertheless, the Stoics made a bold attempt to overcome the difficulties involved in the analysis of motion. Chrysippus, for instance, perceived the intimate connection between time and motion, as is revealed in his "definition" of time, namely, as "the interval of movement with reference to which the measure of speed and slowness is reckoned." He also held that the present moment, the *now*, is given by an infinite sequence of nested time intervals shrinking toward the mathematical "now", a strikingly modern conception.

The problem of the continuum arises also in connection with the *method of exhaustion*. We are told by Archimedes that, using his principle of convergence, Eudoxus successfully proved that the volume of a cone is one third that of the circumscribed cylinder. Archimedes also claims that Democritus originally discovered the result, but was unable to prove it rigorously. The obstacle was that he could see no way of actually building the cone from circular segments, each one of which would differ slightly in area from the two flanking it (the method he had apparently used in discovering the result). The atomist Democritus, with his belief in ultimate finite units, would presumably have understood this "slightly" as entailing a *discrete* difference between the areas of these circular segments, which would produce, not a smooth cone, but instead a ziggurat-like figure with a surface consisting of a series of tiny steps. If, on the other hand, this "slightly" were to be taken to mean "continuously", or "infinitesimally", then the difference between the areas of the segments would seem as

---

[2] See the following chapter for a discussion of infinite sets.

a result to be nonexistent, and one would end up, not with a cone, but a cylinder. Eudoxus later surmounted this difficulty by taking the *limit* of the volumes in a manner essentially similar to the method employed in the integral calculus. This concept of limit is in fact completely in accord with the Stoic conception of the continuum.

The opposition between the continuous and the discrete resurfaced with renewed vigour in the seventeenth century with the emergence of the differential and integral calculus. Here the controversy centred on the concept of *infinitesimal*. According to one school of thought, the infinitesimal was to be regarded as a real, infinitely small, indivisible element of a continuum, similar to the atoms of Democritus, except that now their number was considered to be infinite. Calculation of areas and volumes, i.e., integration, was thought of as summation of an infinite number of these infinitesimal elements. An area, for example, was taken to be the "sum of the lines of which it is formed", as indicated in the diagram below. Thus the continuous was once again reduced to the discrete, but, with the intrusion of the concept of the infinite, in a subtler and more complex way than before.

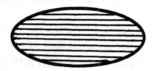

Infinitesimals had a considerable vogue among seventeenth and eighteenth century mathematicians. In the guise of the charmingly named "linelets" and "timelets", they played an essential role in *Isaac Barrow's*[3] (1630–1677) "method for finding tangents by calculation", which appears in his *Lectiones Geometricae* of 1670. As "evanescent quantities" they were instrumental (although later abandoned) in Newton's development of the calculus, and, as "inassignable quantities", in Leibniz's. The *Marquis de l'Hospital* (1661–1704), who in 1696 published the first treatise on the differential calculus (entitled *Analyse des Infiniments Petits pour l'Intelligence des Lignes Courbes*), invokes the concept in postulating that "a curved line may be regarded as being made up of infinitely small straight line segments," and that "one can take as equal two quantities differing by an infinitely small quantity."

However, the conception of infinitesimals as real entities suffered from a certain vagueness and even, on occasion, logical inconsistency. Memorably derided by the philosopher *George Berkeley* (1685–1753) as "ghosts of departed quantities" (and in the twentieth century roundly condemned by Bertrand Russell as "unnecessary, erroneous, and self-contradictory"), this conception of infinitesimal was gradually displaced by the idea—originally suggested by Newton—of the infinitesimal as a *continuous variable* which becomes arbitrarily small. By the start of the nineteenth

---

[3] Barrow is remembered not only for his own outstanding mathematical achievements but also for being the teacher of Newton.

century, when the rigorous theory of limits was in the process of being created, this conception of infinitesimal had been accepted by the majority of mathematicians. A line, for instance, was now understood as consisting not of "points" or "indivisibles", but as the domain of values of a continuous variable, in which separate points are to be considered as locations. At this stage, then, the discrete had given way to the continuous.

But the development of mathematical analysis in the later part of the nineteenth century led mathematicians to demand still greater precision in the theory of continuous variables, and above all in fixing the concept of *real number* as the value of an arbitrary such variable. As a result, in the eighteen seventies a theory was formulated—independently by Dedekind, *Karl Weierstrass* (1815–1897), and *Georg Cantor* (1845–1918)—in which a line is represented as a set of points, and the domain of values of a continuous variable by a set of real numbers. In this scheme of things there was no place for the concept of infinitesimal, which accordingly disappeared for a time. Thus, once again, the continuous was reduced to separate discrete points and the properties of a continuum derived from the structure of its underlying point set. This reduction, underpinned by the development of *set theory*, has led to immense progress in mathematics, and has met with almost universal acceptance by mathematicians.

A new phase in the long contest between the continuous and the discrete has opened in the past few decades with the refounding of the concept of infinitesimal on a solid basis. This has been achieved in two essentially different ways.

First, in the nineteen sixties *Abraham Robinson* (1918–1974), using methods of mathematical logic, created *nonstandard analysis,* an extension of mathematical analysis embracing both "infinitely large" and infinitesimal numbers in which the usual laws of the arithmetic of real numbers continue to hold, an idea which in essence goes back to Leibniz. Here by an infinitely large number is meant one which exceeds every positive integer; the reciprocal of any one of these is infinitesimal in the sense that, while being nonzero, it is smaller than every positive fraction $1/n$. Much of the usefulness of nonstandard analysis stems from the fact that within it every statement of ordinary analysis involving limits has a succinct and highly intuitive translation into the language of infinitesimals. For instance, if we call two numbers x and y *infinitesimally close* when $x - y$ is infinitesimal, we find that the statement

$$\ell \text{ is the limit of } f(x) \text{ as } x \to a$$

is equivalent in meaning to the statement

$$f(a + \varepsilon) \text{ is infinitesimally close to } \ell \text{ for all infinitesimal } \varepsilon \neq 0,$$

and the statement

$$f \text{ is continuous at } a$$

to the statement

f(a + ε) is infinitesimally close to f(a) for all infinitesimal ε ≠ 0.

Nonstandard analysis is presently in a state of rapid development, and has found many applications.

The second development in the refounding of the concept of infinitesimal has been the emergence in the nineteen seventies of *smooth infinitesimal analysis*. Founded on the methods of category theory, this is a rigorous framework of mathematical analysis in which every function between spaces is smooth (i.e., differentiable arbitrarily many times, and so in particular continuous) and in which the use of limits in defining the basic notions of the calculus is replaced by *nilpotent infinitesimals*, that is, of quantities so small (but not actually zero) that some power—most usefully, the square—vanishes[4]. Smooth infinitesimal analysis provides an image of the world in which the continuous is an autonomous notion, not explicable in terms of the discrete.

Thus we see that the opposition between the continuous and the discrete has not ceased to stimulate the development of mathematics.

---

[4] We have already touched on this in the previous chapter: a fuller account will be found in Appendix 3.

CHAPTER 11

THE MATHEMATICS OF THE INFINITE.

THE MOST FAMILIAR EXAMPLE OF AN INFINITE COLLECTION in mathematics is the sequence of positive integers 1, 2, 3,.... . There are many others, for example, the collection of all rational numbers, the collection of all circles in the plane, the collection of all spheres in space, etc. The idea of the infinite is implicit in many mathematical concepts, but it was not until the nineteenth century that the mathematical infinite became the subject of precise analysis. The first steps in this direction were taken by *Bernard Bolzano* (1781–1848) in the first half of that century, but his work went largely unnoticed at that time. The modern theory of the mathematical infinite— *set theory*—was created by Cantor in the latter half of the century. Although set theory initially encountered certain obstacles—which we shall discuss later—it has come to penetrate, and influence decisively, virtually every area of mathematics. It also plays a central role in the logical and philosophical foundations of mathematics.

The basic concept of Cantor's theory is that of *set* or *totality*, which in 1895 he defined in the following way:

> By a *set* we understand every collection to a whole of definite, well-differentiated objects of our intuition or thought.

Another possible definition of a set is that it is a collection of objects, called the *elements or members* of the set, defined by some explicit rule—typically, the possession of a prescribed property—which specifies exactly which objects belong to the collection. All the collections mentioned above are sets in this sense. There is a convenient notation for sets defined in this way. Suppose that P is a property, and write $P(x)$ for the assertion that the object x has the property P. Then

$$\{x: P(x)\}$$

denotes the set whose members are exactly those objects having the property P. In set theory it is also customary to write

$$a \in A \,^{1}$$

---

[1] Here the symbol "∈" is a form of the Greek letter epsilon, the initial letter of the word *esti*, "is". This usage was introduced by Peano.

for *the object a is a member of the set A,* and

$$a \notin A$$

in the contrary case. Clearly, then, for any object a,

$$a \in \{x: P(x)\} \text{ if and only if } P(a).$$

More loosely, if we are given objects $a, b, c, \ldots$, we write $\{a, b, c, \ldots\}$ for the set whose members are $a, b, c, \ldots$.

    A central concept of set theory is that of the *equivalence* of sets, a notion which gives precise expression to the idea of two sets *having the same size.* If the elements of two sets A and B may be paired with each other in such a way that to each element of A there corresponds exactly one element of B, and vice-versa, then, as we recall, the resulting function between A and B is said to be *biunique,* and we shall say that the two sets A and B are *equivalent.* The notion of equivalence for *finite* sets clearly corresponds to the usual concept of *equality of number,* since two finite sets have the same number of elements precisely when the elements of the two sets can be put into biunique correspondence. This is just an extension of the idea of counting, because in counting the elements of a finite set we place them in biunique correspondence with a set of number symbols 1, 2, 3, ..., n. Cantor's idea was to extend the concept of equivalence to *infinite* sets, so as to enable them to be compared in size.

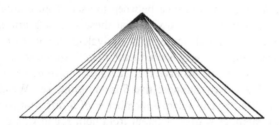

    For infinite sets the idea of equivalence can lead to surprising results. For example, we observe that the sets of points on any pair of line segments are equivalent, even if the line segments are of different lengths. This can be seen by taking two arbitrary line segments and projecting from a point, as in the figure above. By bending one of the segments into a semicircle, a similar procedure, indicated in the figure below, shows that this equivalence obtains even if one of the segments is replaced by the entire line.

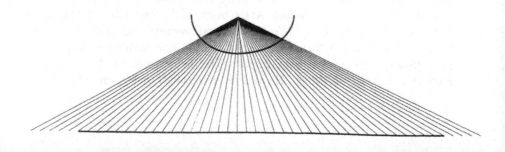

The *even numbers* form a proper subset of the set of *all natural numbers*, and these form a proper subset of the set of all (positive) *rational numbers*. (Here by a *proper* subset of a set S we mean a set consisting of some, but not all, of the objects in S.) Evidently, if a set is *finite*, i.e., if it contains n elements for some natural number n, then it cannot be equivalent to any of its proper subsets, since any such subset would contain at most $n - 1$ elements. But *an infinite set can be equivalent to a proper subset of itself.* For example, the pairing

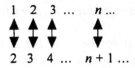

establishes a biunique correspondence between the set of natural numbers and its proper subset of all natural numbers greater than 1, so that the two sets are equivalent. And the pairing

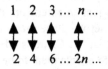

establishes a biunique correspondence between the set of all natural numbers and the proper subset of even numbers, showing that these two sets are equivalent. It is no accident that the set of natural numbers is equivalent to some of its proper subsets, since, as we shall see, this property is *characteristic* of infinite sets.

To prove this we first observe that obviously no *finite* set can be equivalent to any of its proper subsets. Now suppose that S is an *infinite* set. We choose a member s_0 of S. Since S is infinite, we can select a member s_1 of S different from s_0, and then, for the same reason, a member s_2 of S different from both s_0 and s_1, etc. Proceeding in this way, we generate a subset $\{s_0, s_1, s_2, \ldots\} = S'$ of S. Now let $f: S \to S$ be the function defined by setting $f(s_n) = s_{n+1}$ for any natural number n, and $f(x) = x$ if $x \notin S'$. It is easy to check that f is a bijection between S and its proper subset $S - \{s_0\}$. Since s_0 was arbitrary, we see that each infinite set is equivalent to any of its subsets obtained by deleting one element.

The counterintuitive nature of the equivalence between infinite sets and proper subsets is strikingly conveyed by a fable attributed to the German mathematician *David Hilbert* (1862–1943). In Hilbert's tale, he finds himself the manager of a vast hotel, so vast, in fact, that it has an *infinite* number of rooms. Thus the hotel has a first, second,...,n^{th},... room, *ad infinitum*. At the height of the tourist season, Hilbert's hotel is full: each room is occupied. (We are of course assuming the existence of an infinite number of occupants.) Now a newcomer seeking accommodation shows up. "Alas," says Hilbert, "I have not a room to spare." But the newcomer is desperate, and at that point an idea occurs to Hilbert. He asks the occupant of each room to move to the next one; thus the occupant of room 1 is to move to room 2, that of room 2 to room 3, and so on up the line. This leaves the original occupants housed (one to a room, as before), only now the first room is vacant, and the relieved newcomer duly takes possession.

The fable does not end here, however. Suddenly a vast assembly of tourists desiring accommodation descends on the hotel. A quick tally reveals that the assembly is *infinite*, causing Hilbert some consternation. But now another idea occurs to him. This time he requests each occupant to move to the room with *double* the number of the one presently occupied: thus the occupant of room 1 is to proceed to room 2, that of room 2 to room 4, etc. The result is again to leave all the original occupants housed, only now each member of the infinite set of rooms carrying *odd numbers* has become vacant. Thus every newcomer can be accommodated: the first in room 1, the second in room 3, the third in room 5, etc. It is clear that this procedure can be repeated indefinitely, enabling an infinite number of infinite assemblies of tourists to be housed.

Hilbert's fable shows that infinite sets are intriguingly *paradoxical*, but not that they are *contradictory*. Indeed if, for example, the physical universe contains infinitely many stars—a possibility which Newton, for one, was perfectly happy to accept—then it can fill the role of "Hilbert's hotel," with the stars (or orbiting planets) as "rooms".

One of Cantor's first discoveries in his exploration of the infinite was that the set of *rational numbers*—which includes the infinite set of natural numbers and is therefore itself infinite—is actually *equivalent* to the set of natural numbers. At first sight it seems strange that the densely arranged set of rational numbers should be similar to its discretely arranged subset of natural numbers. For, after all, the (positive) rational numbers cannot be discretely arranged in *order of magnitude*: starting with 0 as the first positive rational, we cannot even choose a second "next larger" rational because, for every such choice, there will always be one smaller. Cantor observed that, nevertheless, if we *disregard* the relation of magnitude between successive elements, it *is* possible to arrange all the positive rational numbers in a sequence $r_1, r_2,...$ similar to that of the natural numbers. Such a sequential arrangement of a set of objects is called an *enumeration* of it, and the set is then called *enumerable*. By exhibiting such an enumeration, Cantor established the equivalence of the set of positive rational numbers with the set of natural numbers, since then the correspondence

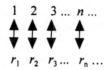

is biunique. How is this enumeration obtained?

Every positive rational number can be written in the form p/q, with p and q natural numbers. We place all these numbers in an array, with p/q in the pth column and q^{th} row: for example, 5/7 is assigned to the fifth column and the seventh row. All the positive rational numbers may now be arranged in a sequence along a continuous "zigzag" line through the array defined as follows. Starting at 1, we proceed to the next entry on the right, obtaining 2 as the second member of the sequence, then diagonally down the the left until the first column is reached at entry 1/2, then vertically down one place to 1/3, diagonally up until the first row is reached again at 3, across to 4, diagonally down to 1/4, and so on, as shown in the figure below. Proceeding along this

1	2	3	4	5	6	7 ...
1/2	2/2	3/2	4/2	5/2	6/2	7/2 ...
1/3	2/3	3/3	4/3	5/3	6/3	7/3 ...
1/4	2/4	3/4	4/4	5/4	6/4	7/4 ...
1/5	2/5	3/5	4/5	5/5	6/5	7/5 ...
1/6	2/6	3/6	4/6	5/6	6/6	7/6 ...
1/7	2/7	3/7	4/7	5/7	6/7	7/7 ...
:	:	:	:	:	:	:

zigzag line we generate a sequence 1, 2, 1/2, 1/3, 2/2, 3, 4, 3/2, 2/3, 1/4, 1/5, 2/4, 3/3, 4/2, 5, ... containing all the positive rationals in the order in which each is encountered along the line. In this sequence we now delete all those numbers p/q for which p and q have a common factor, so that each rational number will appear exactly once and in its simplest form. In this way we obtain a sequence 1, 2, 1/2, 1/3, 3, 4, 3/2, 2/3, 1/4, 1/5, 5, ... in which each rational occurs exactly once. This shows that the set of positive rational numbers is enumerable, and it is easy to deduce from this that the set of all rational numbers—positive, negative, or zero—is also enumerable. For if r_1, r_2, ... is an enumeration of the positive rational numbers, then the sequence 0, r_1, $-r_1$, r_2, $-r_2$, ... is an enumeration of all the rational numbers.

The fact that the rational numbers are enumerable might lead one to suspect that *any* infinite set is enumerable, thus bringing our tour of the infinite to a speedy conclusion. But nothing could be further from the truth. For Cantor also proved the astonishing result that the set of all *real* numbers, rational and irrational, is *not enumerable*. In other words, the totality of real numbers is a radically different, indeed a *larger*, infinity than are the totalities of natural or rational numbers. The idea of Cantor's proof of this fact is to exhibit, for any given enumerated sequence s_1, s_2, ... of real numbers, a new real number which is *outside* it. As an immediate consequence, no given sequence s_1, s_2, ... can enumerate all the real numbers, so that their totality is not enumerable.

Recall that any real number may be regarded as an infinite decimal of the form

$$N. a_1a_2a_3.....$$

where N denotes the integral part and the small letters the digits after the decimal point. Now suppose we are given an enumerated sequence s_1, s_2, s_3, ... of real numbers, with

$$s_1 = N_1. a_1a_2a_3....$$
$$s_2 = N_2. b_1b_2b_3....$$
$$s_3 = N_3. c_1c_2c_3....$$

We proceed to construct, using what has become known as *Cantor's diagonal process*, a new real number which we show does not occur in the given sequence. To do this we first choose a digit a which differs from a_1 and is neither 0 nor 9 (to avoid possible ambiguities which may arise from equalities like 0.999... = 1.000...), then a digit b different from b_2 and again unequal to 0 or 9, similarly c different from c_3 and so on down the "diagonal" of the above array. (For example, we might choose a to be 1 unless a_1 is 1, in which case we choose a to be 2, and similarly down the line for all the digits b, c, d, ...) Now consider the infinite decimal

$$x = 0.abcd....$$

This new real number x then differs from any one of the numbers s_1, s_2, s_3, ...: it cannot be equal to s_1 because the two differ at the first digit after the decimal point; it cannot be equal to s_2 because these two differ at the second digit after the decimal point, and so on. Accordingly x does not occur in the given sequence, so that no given sequence can enumerate the totality of real numbers, which is therefore nonenumerable.

Recalling that the set of real numbers represents the *geometric continuum*, that is, the totality of points on a line, we infer from Cantor's theorem that this latter set is also nonenumerable.

This theorem allows us to draw some further striking conclusions. Let us call a real number *algebraic* if it is a root of an algebraic equation of the form

$$a_n x^n + a_{n-1} x^{n-1} + ... + a_1 x + a_0 = 0,$$

where the a_i are integers. For example, the real numbers $\sqrt{2}$, $\sqrt[3]{5}$ are algebraic. A real number which is not algebraic is called *transcendental*. Do transcendental numbers exist? Cantor's theorem provides the startling answer that, not only do transcendental numbers exist, but there are *nonenumerably many* of them. For it can be shown that the set of algebraic numbers is *enumerable*: indeed, each algebraic number can be correlated with the finite sequence of integer coefficients in the algebraic equation of which it is a root, and a simple extension of the argument used to prove the enumerability of the rationals shows that the set of finite sequences of integers is enumerable. From this it follows that the set of transcendental numbers is nonenumerable. For, if it were enumerable, then the set of all real numbers, as the union of the set of transcendental numbers with the enumerable set of algebraic numbers, would itself be enumerable, contradicting Cantor's theorem.

This conclusion is remarkable in that it shows that "most" real numbers are transcendental *without exhibiting a single one*. Actually the existence of transcendental numbers was established before Cantor's theorem by *J. Liouville* (1809–1882), who provided explicit examples of them, for instance, the infinite decimal

$$0.101001000000100000000000000000000000001.....$$

where the number of zeros between ones increases as $n!$. In 1873 *Charles Hermite* (1822–1901) established the transcendentality of the number e (defined on p. 161) and, as already mentioned, in 1882 Lindemann established the transcendentality of π. In the nineteen thirties A .O. *Gelfond* (1906–1968) showed that numbers like $2^{\sqrt{2}}$ are also transcendental[2].

For set theory the importance of Cantor's theorem stems from the fact that it reveals the presence of at least *two* types of infinity: the enumerable infinity of the set of natural numbers and the nonenumerable infinity of the set of real numbers or the geometric continuum. We shall now extend the use of the term "number" to *arbitrary* —even *infinite*—sets by saying that two equivalent sets, whether they be finite or infinite, have the *same cardinal number*. If A *and* B are finite this of course reduces to the statement that they have the same (natural) number of elements. Let us term *infinite* the cardinal number of an infinite set. We have seen that any infinite set is equivalent to each of its subsets obtained by the deletion of one member, so that adding this one element to the subset does not change its cardinal number. This means that, *for any infinite cardinal number* **k,**

$$\mathbf{k} + 1 = \mathbf{k}.$$

It is the fact that infinite cardinal numbers satisfy this equation that distinguishes them from ordinary integers.

We shall further say that a set A has a *greater cardinal number* than a set B if B is equivalent to a subset of A, but A is not equivalent to B or any of its subsets. Since the set of natural numbers is a subset of the set of real numbers, while, as we have seen, the latter is not equivalent to the former nor to any of its subsets, it follows that the set of real numbers has a greater cardinal number than the set of natural numbers. Cantor went further and established the general fact that, *for any set A, it is possible to produce another set B with a greater cardinal number*. The set B is chosen to be what we have called in Chapter 6 the *power set* of A, that is, the set of all subsets of A, including A itself and the empty subset $\varnothing$ which contains no elements at all. Thus each element of B is *itself* a set, comprising certain elements of A. Now suppose that B were equivalent to A or to some subset of it, that is, there is a biunique correspondence $a| \rightarrow S_a$ between the elements of A (or a subset of it) and the elements of B, i.e. with the subsets of A, where we denote by S_a the subset of A correlated with the element a of A. We show that this is impossible by exhibiting a subset T of A, i.e. an element of B, which *cannot* have any element of A correlated with it. We obtain T as follows. For any element x of A which is correlated with an element S_x of B, we have two possibilities: either x is a member of S_x, or it is not. We define T to be the subset of A consisting of all correlated elements x of A for which x is *not* an element of S_x. We can now show that T is not correlated with any element of A, i.e. that T does not coincide with S_x for any element x of A. For suppose that T and S_a *did* coincide for some specific element a of A. Then, since T consists of all elements x of A for which $x \notin S_x$, it follows, in particular, taking x to be a, that $a \in T$ exactly when $a \notin S_a$. But since T has been assumed to be identical with S_a, $a \in T$ precisely when $a \in S_a$. Thus a is not in S_a exactly when a is in S_a, a

---

[2] In 1934 Gelfond proved the general result that a^b is transcendental whenever $a \neq 0$, 1 is algebraic and b is irrational and algebraic.

contradiction. From this contradiction we conclude that our original assumption was incorrect, and it follows that T cannot coincide with any S_x.

This argument establishes the impossibility of setting up a biunique correspondence between the elements of A, or one of its subsets, and those of its power set B. On the other hand, the pairing $x \mapsto \{x\}$, where, for each x in A, $\{x\}$ denotes the subset of A whose sole element is x, defines a biunique correspondence between A and the subset of B consisting of all one-element subsets of A. It follows from the definition that B has a greater cardinal number than A.

Thus Cantor showed that, just as there is no greatest integer, so too is there no greatest infinite cardinal number, which means that the infinite cardinal numbers form an unbounded ascending sequence. Cantor used the symbol "$\aleph$"—*aleph*, the first letter of the Hebrew alphabet—to denote the members of this sequence, which is then written $\aleph_0, \aleph_1, \aleph_2, \dots$. Its first member, $\aleph_0$, is the cardinal number of the set of integers.

It is a surprising fact that, not only is the cardinal number of the set of points on a line segment independent of its *length*, but the cardinal number of the set of points in a geometric figure is independent of its *number of dimensions*. Thus, for example, we can show that the cardinal number of the set of points in a square is no greater than that of the set of points on one of its sides. This is done by setting up the following correspondence.

If (x, y) is a point of the unit square (i.e., with $0 \le x, y \le 1$), x and y may be written in decimal form as

$$x = 0.a_1a_2a_3a_4\dots$$
$$y = 0.b_1b_2b_3b_4\dots$$

To the point (x, y) we then assign the point

$$z = 0.a_1b_1a_2b_2a_3b_3a_4b_4\dots$$

on the bottom side of the square, obtained by interlacing the two decimals corresponding to x and y. Clearly different points (x, y), (x', y') in the square correspond to different points z, z' on the side, so that we have a biunique correspondence between the set of points in the square and a subset of the set of points on the side. Thus the cardinal number of the set of points in the square does not exceed the cardinal number of the set of points on a side. A similar argument shows that the cardinal number of the set of points in a cube is no greater than that of the set of points on an edge. By slightly refining these arguments it can be shown that the cardinal numbers of these various sets actually *coincide*.

Despite the fact that these results seem to contradict the intuitive idea of dimension, it must be pointed out that the correspondence we have set up is not a *topological* equivalence: it does "tear points apart". Travelling smoothly along the segment from 0 to 1, we find that the corresponding points in the square do not form a smooth curve, but instead form an entirely random pattern. The dimension of a set of points in fact depends primarily on the way its elements are distributed in space, rather than on its cardinal number. This follows from Brouwer's theorem—mentioned in Chapter 8—that the dimension of a region in Euclidean space is a topological invariant.

We have seen that the cardinal number c of the set of real numbers exceeds that of the set of integers. It is natural to ask: by *how much* does the one exceed the other? That is, *how many* cardinal numbers are there (strictly) between that of the set of integers and that of the continuum? The simplest possible answer is: *none*. The conjecture that this is the correct answer is called the *continuum hypothesis*. Since $\aleph_1$ is by definition the next largest cardinal number above $\aleph_0$, the cardinal number of the set of integers, this hypothesis may be succinctly stated in the form

$$c = \aleph_1.$$

Cantor firmly believed in its truth and expended much effort in attempting, without success, to prove it. It was not until the nineteen sixties that *Paul J. Cohen* showed that this hypothesis is in fact *independent* of the principles on which set theory is built (*Kurt Gödel*, 1906–76, having shown in the nineteen thirties that it is consistent with these principles). In fact Cohen showed that c could coincide with virtually *any* cardinal number not less than $\aleph_1$: the principles of set theory simply do not fix an exact value for c in terms of its position in the scale of alephs. Thus the status of the continuum hypothesis in set theory bears a certain resemblance to that of Euclid's fifth postulate in geometry: just as the fifth postulate can be denied so as to produce a noneuclidean geometry, so the continuum hypothesis can be denied so as to produce a "noncantorian" set theory.

Infinite sets, although somewhat paradoxical in character, are, as we have come to realize, not contradictory in themselves. As originally formulated, however, set theory *does* contain contradictions, which result not from admitting infinite totalities *per se,* but rather from countenancing totalities consisting of *all* entities of a certain abstract kind. This is best illustrated by the infamous *Russell paradox*, discovered in 1901 by the philosopher-mathematician *Bertrand Russell* (1872–1970).

Russell's paradox arises in the following way. It starts with the truism that any set is either a member of itself or not. For instance, the set of all cats is not a member of itself since it is not a cat, while the set of all non-cats is a member of itself since it is a non-cat. Now consider the set consisting precisely of all those sets which are *not* members of themselves: call this set *R*. Is *R* a member of itself or not? Suppose it is. Then it must satisfy the defining condition for inclusion in *R*, i.e. it must *not* be a member of itself. Conversely, suppose it is not a member of itself. Then it *fails* to satisfy the defining condition for inclusion in *R*, that is, it *must be* a member of itself. We have thus arrived at the unsettling, indeed contradictory, conclusion that *R* is a member of itself precisely when it is not. We note that whether *R* is finite or infinite is irrelevant; the argument depends solely on the defining property for membership in *R*.

Russell's paradox also appears when we consider such curious entities as, for instance, the bibliography of all bibliographies that fail to list themselves: such a bibliography would, if it existed, list itself precisely when it does not. In this case, however, we may simply infer that the entity in question does not exist, a conclusion we cannot draw in the case of the Russell set R without bringing into question the very basis on which sets have been introduced.

Russell's paradox has a purely *linguistic* counterpart known as the *Grelling-Nelson paradox*. Call an (English) adjective *autological* if it is true of itself and *heterological* if not. For instance, the adjectives "polysyllabic", "English" are autological, and "palindromic", "French" are heterological. Now consider the adjective "heterological". Is it autological or not? A moment's thought reveals that it *is* precisely when it *is not*..

Another principle of set theory whose enunciation (in 1904 by *Ernst Zermelo*, 1871–1953) occasioned much dispute is the so-called *axiom of choice*. In its simplest form, the axiom asserts that, if we are given any collection S of sets, each of which has at least one member, then there is a set M containing exactly one element from each set in S.[3] No difficulty is encountered in assembling M when there are only finitely many sets in S, or if S is infinite, but we possess a *definite rule* for choosing a member from each set in S. The problem arises when S contains infinitely many sets, but we have *no rule* for selecting a member from each: in this situation, how can the procedure be justified of making infinitely, perhaps even nonenumerably, many arbitrary choices, and forming a set from the result?

The difficulty here is well illustrated by a Russellian anecdote. A millionaire possesses an infinite number of pairs of shoes, and an infinite number of pairs of socks. One day, in a fit of eccentricity, he summons his valet and asks him to select one shoe from each pair. When the valet, accustomed to receiving precise instructions, asks for details as to how to perform the selection, the millionaire suggests that the left shoe be chosen from each pair. Next day the millionaire proposes to the valet that he select one *sock* from each pair. When asked as to how *this* operation is to be carried out, the millionaire is at a loss for a reply, since, unlike shoes, there is no intrinsic way of distinguishing one sock of a pair from the other. In other words, the selection of the socks must be truly arbitrary.

One curious consequence of the axiom of choice is the *paradoxical decomposition of the sphere*, formulated in 1924 by the Polish mathematicians *Stefan Banach* (1892–1945) and *Alfred Tarski* (1902–1983). In one form, it asserts that a solid sphere can be cut up into finitely many (later shown to be reducible to five!) pieces which can themselves be reassembled exactly to form *two* solid spheres, *each* of the same size as the original. Another version of this "paradox" is that, given any two solid spheres, either one of them can be cut up into finitely many pieces which can be reassembled to form a solid sphere of the same size as the other. Thus, for example, a sphere the size of the sun can be cut up and reassembled so as to form a sphere the size of a pea! Of course, the phrase "can be cut up" here is to be taken in a metaphorical, not practical, sense; but this does not detract from the counterintuitiveness of these results. Strange as they are, however, unlike Russell's paradox, they do not constitute outright contradictions: sphere decompositions become possible in set theory only

---

[3] Implicit use of the axiom of choice was made in our proof that every infinite set is equivalent to a proper subset of itself. It can be shown that this use is essential.

because continuous geometric objects have been analyzed into discrete sets of points
which can then be rearranged in an arbitrary manner.

The perplexities surrounding the emergence of set theory are collectively
designated by historians of mathematics as the *third crisis* in the foundations of
mathematics (the two previous being the Pythagorean discovery of incommensurables,
and the shaky state of the foundations of the calculus in the seventeenth and eighteenth
centuries). Attempts to resolve this crisis took several different forms, all of which
involved subtle analysis of the nature of mathematical concepts and reasoning—topics
in *mathematical logic* and the *philosophy of mathematics* which we take up in the final
chapter.

The purely technical difficulties in set theory were overcome when, in the first
few decades of this century, it was *axiomatized* (by Zermelo and others) in such a way
as to circumvent contradictions such as Russell's paradox by suitably restricting the
formation rules for sets. Any residual doubts concerning the acceptability of the axiom
of choice were dispelled in 1938 when Gödel established its consistency with respect to
the remaining axioms of set theory. These developments enabled the majority of
mathematicians to accept set theory as providing an adequate foundation for their work.
Mathematicians find set theory acceptable not solely for the practical reason that it
enables mathematics to be *done,* but also because it is consonant with the unspoken
belief of the majority that mathematical objects actually *exist* in some sense and
mathematical theorems express truths about these objects. This is a version of the
philosophical doctrine of *Realism,* also termed, with less accuracy, *Platonism.*

Hermann Weyl has written:

'Mathematizing' may well be a human creative activity, like music, the products of which not
only in form but also in substance are conditioned by the decisions of history and therefore defy
complete objective rationalization.

This is a perceptive observation; nevertheless, the attempt to explicate the mathematical
infinite, and so to grasp the ultimate nature of mathematics, has borne much fruit and
will no doubt long continue to be a source of inspiration.

CHAPTER 12

THE PHILOSOPHY OF MATHEMATICS.

THE CLOSE CONNECTION BETWEEN mathematics and philosophy has long been recognized by practitioners of both disciplines. The apparent timelessness of mathematical truth, the exactness and objective nature of its concepts, its applicability to the phenomena of the empirical world—explicating such facts presents philosophy with some of its subtlest problems. In this final chapter we discuss some of the attempts made by philosophers and mathematicians to explain the nature of mathematics. We begin with a brief presentation of the views of four major classical philosophers: *Plato, Aristotle, Leibniz,* and *Kant.* We conclude with a more detailed discussion of the three "schools" of mathematical philosophy which have emerged in the twentieth century: *Logicism, Formalism,* and *Intuitionism.*

Classical Views on the Nature of Mathematics.

Plato (*c.*428–347 B.C.) included mathematical entities—numbers and the objects of pure geometry such as points, lines, and circles—among the well-defined, independently existing eternal objects he called *Forms.* It is the fact that mathematical statements refer to these definite Forms that enables such statements to be true (or false). Mathematical statements about the empirical world are true to the extent that sensible objects resemble or manifest the corresponding Forms. Plato considered mathematics not as an idealization of aspects of the empirical world, but rather as *a direct description of reality,* that is, the world of Forms as apprehended by reason.

Plato's pupil and philosophical successor *Aristotle* (384–322 B.C.), on the other hand, rejected the notion of Forms being separate from empirical objects, and maintained instead that *the Forms constitute parts of objects.* Forms are grasped by the mind through a process of *abstraction* from sensible objects, but they do not thereby attain an autonomous existence detached from these latter. Mathematics arises from this process of abstraction; its subject matter is the body of idealizations engendered by this process; and mathematical rigour arises directly from the simplicity of the properties of these idealizations. Aristotle rejected the concept of *actual* (or completed) *infinity,* admitting only *potential infinity,* to wit, that of a totality which, while finite at any given time, grows beyond any preassigned bound, e.g. the sequence of natural numbers or the process of continually dividing a line.

Leibniz divided all true propositions, including those of mathematics, into two types: *truths of fact*, and *truths of reason*, also known as *contingent* and *analytic* truths, respectively. According to Leibniz, true mathematical propositions are truths of reason and their truth is therefore just logical truth: their denial would be logically impossible. Mathematical propositions do not have a special "mathematical" content—as they did for Plato and Aristotle—and so true mathematical propositions are true in all possible worlds, that is, they are *necessarily* true.[1] Leibniz attached particular importance to the *symbolic* aspects of mathematical reasoning. His program of developing a *characteristica universalis* centered around the idea of devising a method of representing thoughts by means of arrangements of characters and signs in such a way that relations among thoughts are reflected by similar relations among their representing signs.

Immanuel Kant (1724–1804) introduced a new classification of (true) propositions: *analytic*, and nonanalytic, or *synthetic*, which he further subdivided into empirical, or *a posteriori*, and nonempirical, or *a priori*. Synthetic a priori propositions are not dependent on sense perception, but are necessarily true in the sense that, if *any* propositions about the empirical world are true, *they* must be true. According to Kant, mathematical propositions are synthetic a priori because they ultimately involve reference to *space and time*. Kant attached particular importance to the idea of *a priori construction* of mathematical objects. He distinguishes sharply between mathematical *concepts* which, like noneuclidean geometries, are merely internally consistent, and mathematical *objects* whose construction is made possible by the fact that perceptual space and time have a certain inherent structure. Thus, on this reckoning, $2 + 3 = 5$ is to be regarded ultimately as an assertion about a certain construction, carried out in time and space, involving the succession and collection of units. The *logical possibility* of an arithmetic in which $2 + 3 \neq 5$ is not denied; it is only asserted that the correctness of such an arithmetic would be incompatible with the structure of perceptual space and time. So for Kant the propositions of pure arithmetic and geometry are necessary, but *synthetic a priori*. *Synthetic,* because they are ultimately about the structure of space and time, revealed through the objects that can be constructed there. And *a priori* because the structure of space and time provides the universal preconditions rendering possible the perception of such objects. On this reckoning, *pure mathematics* is the analysis of the structure of pure space and time, free from empirical material, and *applied mathematics* is the analysis of the structure of space and time, augmented by empirical material. Like Aristotle, Kant distinguishes between *potential* and *actual infinity*. However, Kant does not regard actual infinity as being a logical impossibility, but rather, like noneuclidean geometry, as an *idea of reason*, internally consistent but neither perceptible nor constructible.

---

[1] On the other hand, empirical propositions containing mathematical terms such as 2 *cats* $+ 3$ *cats* $= 5$ *cats* are true because they hold in the *actual* world, and, according to Leibniz, this is the case only because the actual world is the "best possible" one. Thus, despite the fact that $2 + 3 = 5$ is true in all possible worlds, 2 *cats* $+ 3$ *cats* $= 5$ *cats* could be false in some world.

Logicism

The Greeks had developed mathematics as a rigorous demonstrative science, in which geometry occupied central stage. But they lacked an abstract conception of *number*: this in fact only began to emerge in the Middle Ages under the stimulus of Indian and Arabic mathematicians, who brought about the liberation of the number concept from the dominion of geometry. The seventeenth century witnessed two decisive innovations which mark the birth of modern mathematics. The first of these was introduced by Descartes and Fermat, who, through their invention of coordinate geometry, succeeded in correlating the then essentially separate domains of algebra and geometry, so paving the way for the emergence of modern mathematical analysis. The second great innovation was, of course, the development of the infinitesimal calculus by Leibniz and Newton.

However, a price had to be paid for these achievements. In fact, they led to a considerable diminution of the deductive rigour on which the certainty of Greek mathematics had rested. This was especially true in the calculus, where the rapid development of spectacularly successful new techniques for solving previously intractable problems excited the imagination of mathematicians to such an extent that they frequently threw logical caution to the winds and allowed themselves to be carried away by the spirit of adventure. A key element in these techniques was the concept of *infinitesimal quantity* which, although of immense fertility, was logically somewhat dubious. By the end of the eighteenth century a somewhat more circumspect attitude to the cavalier use of these techniques had begun to make its appearance, and in the nineteenth century serious steps began to be taken to restore the tarnished rigour of mathematical demonstration. The situation (in 1884) was summed up by Frege in a passage from his *Foundations of Arithmetic:*

> After deserting for a time the old Euclidean standards of rigour, mathematics is now returning to them, and even making efforts to go beyond them. In arithmetic, it has been the tradition to reason less strictly than in geometry. The discovery of higher analysis only served to confirm this tendency; for considerable, almost insuperable, difficulties stood in the way of any rigorous treatment of these subjects, while at the same time small reward seemed likely for the efforts expended in overcoming them. Later developments, however, have shown more and more clearly that in mathematics a mere moral conviction, supported by a mass of succesful applications, is not good enough. Proof is now demanded of many things that formerly passed as self-evident. Again and again the limits to the validity of a proposition have been in this way established for the first time. The concepts of function, of continuity, of limit and of infinity have been shown to stand in need of sharper definition. Negative and irrational numbers, which had long since been admitted into science, have had to admit to a closer scrutiny of their credentials. In all directions these same ideals can be seen at work—rigour of proof, precise delimitation of extent of validity, and as a means to this, sharp definition of concepts.

Both Frege and Dedekind were concerned to supply mathematics with rigorous definitions. They believed that the central concepts of mathematics were ultimately *logical* in nature, and, like Leibniz, that truths about these concepts should be established by purely logical means. For instance, Dedekind asserts (in the Preface to his *The Nature and Meaning of Numbers*, 1888) that

> I consider the number concept [to be] entirely independent of the notions or intuitions of space
> and time ... an immediate result from the laws of thought.

Thus, if we make the traditional identification of logic with the laws of thought, Dedekind is what we would now call a *logicist* in his attitude toward the nature of mathematics. Dedekind's "logicism" embraced *all* mathematical concepts: the concepts of number—natural, rational, real, complex—and geometric concepts such as continuity[2]. As a practicing mathematician Dedekind brought a certain latitude to the conception of what was to count as a "logical" notion—a law of thought—as is witnessed by his remark that

> ... we are led to consider the ability of the mind to relate things to things, to let a thing
> correspond to a thing, or to represent a thing by a thing, an ability without which no thinking is
> possible.[3]

Thus Dedekind was not particularly concerned with providing precise formulation of the logical principles supporting his reasoning, believing that reference to self-evident "laws of thinking" would suffice. Dedekind's logicism was accordingly of a less thoroughgoing and painstaking nature than that of his contemporary Frege, whose name, together with Bertrand Russell's, is virtually synonymous with logicism. In his logical analysis of the concept of number, Frege undertook to fashion in exacting detail the symbolic language within which his analysis was to be presented.

Frege's analysis is presented in three works:

Begriffsschrift (1879): Concept-Script, a symbolic language of pure thought modelled on the language of arithmetic.

Grundlagen (1884): The Foundation of Arithmetic, a logico-mathematical investigation into the concept of number.

Grundgesetze (1893, 1903): Fundamental Laws of Arithmetic, derived by means of concept-script.

In the *Grundgesetze* Frege refines and enlarges the symbolic language first introduced in the *Begriffsschrift* so as to undertake, in full formal detail, the analysis of the concept of number, and the derivation of the fundamental laws of arithmetic. The logical universe of *Grundgesetze* comprises two sorts of entity: *functions*, and *objects*. Any function f associates with each value ξ of its argument an object $f(\xi)$: if this object is always one of the two *truth values* 0 (false) or 1 (true), then f is called a *concept* or *propositional function*, and when $f(\xi) = 1$ we say that ξ *falls under* the concept f. If two functions f and g assign the same objects to all possible values of their arguments, we should naturally say that they have the same *course of values*; if f and g are concepts, we would say that they both have the same *extension*. Frege's decisive step in

---

[2] In fact, it was the imprecision surrounding the concept of continuity that impelled him to embark on the program of critical analysis of mathematical concepts.

[3] This idea of *correspondence* or *functionality*, taken by Dedekind as fundamental, is in fact the central concept of *category theory* (see Chapter 6).

the *Grundgesetze* was to introduce a new kind of object expression—which we shall write as $f^\wedge$—to symbolize the course of values of f and to lay down as a basic principle the assertion

$$f^\wedge = g^\wedge \leftrightarrow \forall \xi \; [f(\xi) = g(\xi)].^4 \tag{1}$$

Confining attention to concepts, this may be taken as asserting that *two concepts have the same extension exactly when the same entities fall under them.*

The notion of the extension of a concept underpins Frege's definition of number, which in the *Grundlagen* he had argued persuasively should be taken as a measure of the size of a concept's extension[5]. As already mentioned in Chapter 3, he introduced the term *equinumerous* for the relation between two concepts that obtains when the fields of entities falling under each can be put in biunique correspondence. He then defined cardinal number by stipulating that the cardinal number of a concept F is the extension of the concept *equinumerous with the concept F*. In this way a number is associated with a *second-order* concept—a concept about concepts. Thus, if we write $v(F)$ for the cardinal number of F so defined, and $F \approx G$ for *the concept F is equinumerous with the concept G*, then it follows from (1) that

$$v(F) = v(G) \leftrightarrow F \approx G. \tag{2}$$

And then the natural numbers can be defined as the cardinal numbers of the following concepts:

$$N_0 : x \neq x \qquad\qquad 0 = v(N_0)$$

$$N_1 : x = 0 \qquad\qquad 1 = v(N_1)$$

$$N_2 : x = 0 \vee x = 1 \qquad\qquad 2 = v(N_2)$$

etc.

In a technical *tour-de-force* Frege established that the natural numbers so defined satisfy the usual principles expected of them.

Unfortunately, in 1902 Frege learned of *Russell's paradox*, which can be derived from his principle (1) and shows it to be *inconsistent*. Russell's paradox, as formulated for sets or classes in the previous chapter, can be seen to be attendant upon the usual supposition that *any property determines a unique class*, to wit, the class of

---

[4] Here and in the sequel we employ the logical operators introduced at the beginning of Appendix 2. Thus "$\forall$" stands for "for every", "$\exists$" stands for "there exists", "$\neg$" for "not"," $\wedge$" for 'and", "$\vee$" for "or", "$\rightarrow$" for "implies" and "$\leftrightarrow$" for "is equivalent to".

[5] It is helpful to think of the extension of a concept as the class of all entities that fall under it, so that, for example, the extension of the concept *red* is the class of all red objects. However, it is by no means necessary to identify extensions with classes; all that needs to be known about extensions is that they are objects satisfying (1).

all objects possessing that property (its "extension"). To derive the paradox in Frege's system, classes are replaced by Frege's extensions: we define the concept R by

$$R(x) \leftrightarrow \exists F[x = F^\wedge \wedge \neg F(x)]$$

(in words: *x falls under the concept R exactly when x is the extension of some concept under which it does not fall*). Now write r for the extension of R, i.e.,

$$r = R^\wedge.$$

Then

$$R(r) \leftrightarrow \exists F[r = F^\wedge \wedge \neg F(r)]. \tag{3}$$

Now suppose that $R(r)$ holds. Then, for some concept F,

$$r = F^\wedge \wedge \neg F(r).$$

But then

$$F^\wedge = r = R^\wedge,$$

and so we deduce from (1) that

$$\forall x[F(x) \leftrightarrow R(x)].$$

Since $\neg F(r)$, it follows that $\neg R(r)$. We conclude that

$$R(r) \rightarrow \neg R(r).$$

Conversely, assume $\neg R(r)$. Then

$$r = R^\wedge \wedge \neg R(r),$$

and so *a fortiori*

$$\exists F[r = F^\wedge \wedge \neg F(r)].$$

It now follows from the definition of R that $\neg R(r)$. Thus we have shown that

$$\neg R(r) \rightarrow R(r).$$

We conclude that Frege's principle (1) yields the contradiction

$$R(r) \leftrightarrow \neg R(r).$$

Thus Frege's system in the *Grundgesetze* is, as it stands, inconsistent. Later investigations, however, have established that the definition of the natural numbers and the derivation of the basic laws of arithmetic can be salvaged by suitably restricting (1) so that it becomes consistent, leaving the remainder of the system intact. In fact it is only necessary to make the (consistent) assumption that the extensions of a certain special type of concept—the *numerical concepts*[6]—satisfy (1). Alternatively, one can abandon extensions altogether and instead take the cardinal number $v(F)$ as a primitive notion, governed by equivalence (2). In either case the whole of Frege's derivation of the basic laws of arithmetic can be recovered.

Where does all this leave Frege's (and Dedekind's) claim that arithmetic can be derived from logic? Both established beyond dispute that arithmetic can be formally or logically derived from principles which involve no explicit reference to spatiotemporal intuitions. In Frege's case the key principle involved is that certain concepts have extensions satisfying (1). But although this principle involves no reference to spatiotemporal intuition, it can hardly be claimed to be of a purely logical nature. For it is an *existential* assertion and one can presumably conceive of a world devoid of the objects ("extensions") whose existence is asserted. It thus seems fair to say that, while Frege (and Dedekind) did succeed in showing that the concept and properties of number are "logical" in the sense of being independent of spatiotemporal intuition, they did not (and it would appear could not) succeed in showing that these are "logical" in the stronger Leibnizian sense of holding in every possible world.

The logicism of *Bertrand Russell* was in certain respects even more radical than that of Frege, and closer to the views of Leibniz. In *The Principles of Mathematics* (1903) he asserts that mathematics and logic are *identical*. To be precise, he proclaims at the beginning of this remarkable work that

> Pure mathematics is the class of all propositions of the form "p implies q" where p and q are propositions ... and neither p nor q contains any constants except logical constants.[7]

The monumental, and formidably recondite[8] *Principia Mathematica*, written during 1910 –1913 in collaboration with *Alfred North Whitehead* (1861–1947), contains a complete system of pure mathematics, based on what were intended to be

---

[6] A numerical concept is one expressing equinumerosity with some given concept.

[7] Thus at the time this was asserted Russell was what could be described as an "implicational logicist".

[8] One may get an idea of just how difficult this work is by quoting the following extract from a review of it in a 1911 number of the London magazine *The Spectator:*

> It is easy to picture the dismay of the innocent person who out of curiosity looks into the later part of the book. He would come upon whole pages without a single word of English below the headline; he would see, instead, scattered in wild profusion, disconnected Greek and Roman letters of every size interspersed with brackets and dots and inverted commas, with arrows and exclamation marks standing on their heads, and with even more fantastic signs for which he would with difficulty so much as find names.

purely logical principles, and formulated within a precise symbolic language. A central concern of *Principia Mathematica* was to avoid the so-called *vicious circle paradoxes,* such as those of Russell and Grelling-Nelson—mentioned in the previous chapter— which had come to trouble mathematicians concerned with the ultimate soundness of their discipline. Another is *Berry's paradox,* in one form of which we consider the phrase *the least integer not definable in less than eleven words.* This phrase defines, in less than eleven words (ten, actually), an integer which satisfies the condition stated, that is, of not being definable in less than eleven words. This is plainly self-contradictory.

If we examine these paradoxes closely, we find that in each case a term is defined by means of an implicit reference to a certain class or domain which contains the term in question, thereby generating a vicious circle. Thus, in Russell's paradox, the defined entity, that is, the class R of all classes not members of themselves is obtained by singling out, from the class V of all classes *simpliciter,* those that are not members of themselves. That is, R is defined in terms of V, but since R is a member of V, V cannot be obtained without being given R in advance. Similarly, in the Grelling-Nelson paradox, the definition of the adjective *heterological* involves considering the concept *adjective* under which *heterological* itself falls. And in the Berry paradox, the term *the least integer not definable in less than eleven words* involves reference to the class of all English phrases, including the phrase defining the term in question.

Russell's solution to these problems was to adopt what he called the *vicious circle principle* which he formulated succinctly as: *whatever involves* all *of a collection must not be one of a collection.* This injunction has the effect of excluding, not just self-contradictory entities of the above sort, but *all* entities whose definition is in some way circular, even those, such as the class of all classes which are members of themselves, the adjective "autological", or the least integer definable in less than eleven words, the supposition of whose existence does not appear to lead to contradiction.[9]

The vicious circle principle suggests the idea of arranging classes or concepts (propositional functions) into distinct *types* or *levels,* so that, for instance, any class may only contain classes (or individuals) of lower level as members[10], and a propositional function can have only (objects or) functions of lower level as possible arguments. Under the constraints imposed by this theory, one can no longer form the class of all possible classes as such, but only the class of all classes of a given level. The resulting class must then be of a higher type than each of its members, and so cannot be a member of itself. Thus Russell's paradox cannot arise. The Grelling-Nelson paradox is blocked because the property of heterologicality, which involves self-application, is inadmissible.

Unfortunately, however, this simple theory of types does not circumvent

---

[9] The self-contradictory nature of the "paradoxical" entities we have described derives as much from the occurrence of *negation* in their definitions as it does from the circularity of those definitions.

[10] The idea of stratifying classes into types had also occurred to Russell in connection with his analysis of classes as genuine *pluralities,* as opposed to *unities.* On this reckoning, one starts with individual objects (lowest type), pluralities of these comprise the entities of next highest type, pluralities of these pluralities the entities of next highest type, etc. Thus the evident distinction between individuals and pluralities is "projected upwards" to produce a hierarchy of types.

paradoxes such as Berry's, because in these cases the defined entity is clearly of the same level as the entities involved in its definition. To avoid paradoxes of this kind Russell was therefore compelled to introduce a further "horizontal" subdivision of the totality of entities at each level, into what he called *orders*, and in which the *mode of definition* of these entities is taken into account. The whole apparatus of types and orders is called the *ramified theory of types* and forms the backbone of the formal system of *Principia Mathematica*.

To convey a rough idea of how Russell conceived of orders, let us confine attention to propositional functions taking only individuals (type 0) as arguments. Any such function which can be defined without application of quantifiers[11] to any variables other than individual variables is said to be of *first order*. For example, the propositional function *everybody loves x* is of first order. Then *second order* functions are those whose definition involves application of quantifiers to nothing more than individual and first order variables, and similarly for third, fourth,..., n^{th} order functions. Thus *x has all the first-order qualities that make a great philosopher* represents a function of second order and first type.

Distinguishing the order of functions enables paradoxes such as Berry's to be dealt with. There the word *definable* is incorrectly taken to cover not only definitions in the usual sense, that is, those in which no functions occur, but also definitions involving functions of all orders. We must instead insist on specifying the orders of all functions figuring in these definitions. Thus, in place of the now illegitimate *the least integer not definable in less than eleven words* we consider *the least integer not definable in terms of functions of order* n *in less than eighteen words*. This integer is then indeed *not* definable in terms of functions of order n in less than eighteen words, but *is* definable in terms of functions of order $n+1$ in less than eighteen words. There is no conflict here.

While the ramified theory of types circumvents all known paradoxes (and can in fact be proved consistent from some modest assumptions), it turns out to be too weak a system to support unaided the development of mathematics. To begin with, one cannot prove within it that there is an infinity of natural numbers, or indeed that each natural number has a distinct successor. To overcome this deficiency Russell was compelled to introduce an *axiom of infinity*, to wit, that there exists a level containing infinitely many entities. As Russell admitted, however, this can hardly be considered a principle of logic, since it is certainly possible to conceive of circumstances in which it might be false. In any case, even augmented by the axiom of infinity, the ramified theory of types proves inadequate for the development of the basic theory of the real numbers. For instance, the theorem that every bounded set of real numbers has a least upper bound, upon which the whole of mathematical analysis rests, is not derivable without further *ad hoc* strengthening of the theory, this time by the assumption of the so-called *axiom of reducibility*. This asserts that any propositional function of any order is equivalent to one of first order in the sense that the same entities fall under them. Again, this principle can hardly be claimed to be a fact of logic.

---

[11]Here by a *quantifier* we mean a an expression of the form "for every" (∀) or "there exists" (∃).

Various attempts have been made to dispense with the axiom of reducibility, notably that of *Frank Ramsey* (1903–1930). His idea was to render the whole apparatus of orders superfluous by eliminating quantifiers in definitions. Thus he proposed that a universal quantifier be regarded as indicating a conjunction, and an existential quantifier a disjunction[12], even though it may be impossible in practice to write out the resulting expressions in full. On this reckoning, then, the statement *Citizen Kane has all the qualities that make a great film* would be taken as an abbreviation for something like *Citizen Kane is a film, brilliantly directed, superbly photographed, outstandingly performed, excellently scripted, etc.* For Ramsey, the distinction of orders of functions is just a complication imposed by the structure of our language and not, unlike the hierarchy of types, something inherent in the way things are. For these reasons he believed that the simple theory of types would provide an adequate foundation for mathematics.

What is the upshot of all this for Russell's logicism? There is no doubt that Russell and Whitehead succeeded in showing that mathematics can be derived within the ramified theory of types from the axioms of infinity and reducibility. This is indeed no mean achievement, but, as Russell admitted, the axioms of infinity and reducibility seem to be at best contingent truths. In any case it seems strange to have to base the truth of mathematical assertions on the proviso that there are infinitely many individuals in the world. Thus, like Frege's, Russell's attempted reduction of mathematics to logic contains an irreducible mathematical residue.

Formalism

In 1899 Hilbert published his epoch-making work *Grundlagen der Geometrie* ("Foundations of Geometry"). Without introducing any special symbolism, in this work Hilbert formulates an absolutely rigorous axiomatic treatment of Euclidean geometry, revealing the hidden assumptions, and bridging the logical gaps, in traditional accounts of the subject. He also establishes the *consistency* of his axiomatic system by showing that they can be interpreted (or as we say, possess a *model*) in the system of real numbers. Another important property of the axioms he demonstrated is their *categoricity*, that is, the fact that, up to isomorphism they have exactly *one* model, namely, the usual 3-dimensional space of real number triples. Although in this work Hilbert was attempting to show that geometry is entirely self-sufficient *as a deductive system*[13], he nevertheless thought, as did Kant, that geometry is ultimately *the logical analysis of our intuition of space*. This can be seen from the fact that as an epigraph for his book he quotes Kant's famous remark from the *Critique of Pure Reason*:

Human knowledge begins with intuitions, goes from there to concepts, and ends with ideas.

---

[12]Thus, if a domain of discourse D comprises entities $a,b,c,...,$ then *for every x in D*, $P(x)$ is construed to mean $P(a)$ *and* $P(b)$ and $P(c)$ *and...*, and *there exists x in D such that* $P(x)$ to mean $P(a)$ *or* $P(b)$ *or* $P(c)$ *or* ...

[13]In this connection one recalls his famous remark: *one must be able to say at all times, instead of points, lines and planes—tables, chairs, and beer mugs.*

The great success of the method Hilbert had developed to analyze the deductive system of Euclidean geometry—we might call it the *rigorized axiomatic method*, or the *metamathematical method*—emboldened him to attempt later to apply it to pure mathematics as a whole, thereby securing what he hoped to be perfect rigour for all of mathematics. To this end Hilbert elaborated a subtle philosophy of mathematics, later to become known as *formalism*, which differs in certain important respects from the logicism of Frege and Russell and manifests certain Kantian features. Its flavour is well captured by the following quotation from an address he made in 1927:

> No more than any other science can mathematics be founded on logic alone; rather, as a condition for the use of logical inferences and the performance of logical operations, something must already be given to us in our faculty of representation, certain extralogical concrete objects that are intuitively present as immediate experience prior to all thought. If logical inference is to be reliable, it must be possible to survey these objects completely in all their parts, and the fact that they occur, that they differ from one another, and that they follow each other, or are concatenated, is immediately given intuitively, together with the objects, as something that can neither be reduced to anything else, nor requires reduction. This is the basic philosophical position that I regard as requisite for mathematics and, in general, for all scientific thinking, understanding, and communication. And in mathematics, in particular, what we consider is the concrete signs themselves, whose shape, according to the conception we have adopted, is immediately clear and recognizable. This is the very least that must be presupposed, no scientific thinker can dispense with it, and therefore everyone must maintain it, consciously or not.

Thus, at bottom, Hilbert, like Kant, wanted to ground mathematics on the description of concrete spatiotemporal configurations, only Hilbert restricts these configurations to *concrete signs* (such as inscriptions on paper). No inconsistencies can arise within the realm of concrete signs, since precise descriptions of concrete objects are always mutually compatible. In particular, within the mathematics of concrete signs, actual infinity cannot generate inconsistencies since, as for Kant, this concept cannot describe any concrete object. On this reckoning, the soundness of mathematics thus issues ultimately, not from a *logical* source, but from a *concrete* one, in much the same way as the consistency of truly reported empirical statements is guaranteed by the concreteness of the external world.

Yet Hilbert also thought that adopting this position would not require the abandonment of the infinitistic mathematics of Cantor and others which had emerged in the nineteenth century and which had enabled mathematics to make such spectacular strides. He accordingly set himself the task of accommodating infinitistic mathematics within a mathematics restricted to the deployment of finite concrete objects. Thus *Hilbert's program*, as it came to be called, had as its aim the provision of a new foundation for mathematics not by reducing it to logic, but instead by *representing its essential form within the realm of concrete symbols*. As the quotation above indicates, Hilbert considered that, in the last analysis, the completely reliable, irreducibly self-evident constituents of mathematics are *finitistic,* that is, concerned just with finite manipulation of surveyable domains of concrete objects, in particular, mathematical symbols presented as marks on paper. Mathematical propositions referring only to concrete objects in this sense he called *real*, or *concrete*, propositions, and all other mathematical propositions he considered as possessing an *ideal*, or *abstract* character. Thus, for example, $2 + 2 = 4$ would count as a real proposition, while *there exists an odd perfect number* would count as an ideal one.

Hilbert viewed ideal propositions as akin to the ideal lines and points "at infinity" of projective geometry. Just as the use of these does not violate any truths of the "concrete" geometry of the usual Cartesian plane, so he hoped to show that the use of ideal propositions—in particular, those of Cantor's set theory—would never lead to falsehoods among the real propositions, that, in other words, such use *would never contradict any self-evident fact about concrete objects.* Establishing this by strictly concrete, and so unimpeachable means was the central aim of Hilbert's program. In short, its objective was to prove classical mathematical reasoning *consistent.* With the attainment of this goal, mathematicians would be free to roam unconstrained within "Cantor's Paradise" (in Hilbert's memorable phrase[14]). This was to be achieved by setting it out as a purely formal system of symbols, devoid of meaning[15], and then showing that no proof in the system can lead to a false assertion, e.g. $0 = 1$. This, in turn, was to be done by employing the *metamathematical* technique of replacing each abstract classical proof of a real proposition by a concrete, finitistic proof. Since, plainly, there can be no concrete proof of the real proposition $0 = 1$, there can be no classical proof of this proposition either, and so classical mathematical reasoning is consistent.

In 1931, however, Gödel rocked Hilbert's program by demonstrating, through his celebrated *Incompleteness Theorems*, that *there would always be real propositions provable by ideal means which cannot be proved by concrete means.* He achieved this by means of an ingenious modification of the ancient *Liar paradox.* To obtain the liar paradox in its most transparent form, one considers the sentence *this sentence is false.* Calling this sentence *A,* it is clear that *A* is true if and only if it is false, that is, *A asserts its own falsehood.* Now Gödel showed that, if in *A* one replaces the word *false* by the phrase *not concretely provable,* then the resulting statement *B* is *true* —i.e., provable by ideal means—but *not concretely provable.* This is so because, as is easily seen, *B* actually asserts its own concrete unprovability in just the same way as *A* asserts its own falsehood. And by extending these arguments Gödel also succeeded in showing that the *consistency of arithmetic* cannot be proved by concrete means.[16]

Accordingly there seems to be no doubt that Hilbert's program for establishing the consistency of mathematics (and in particular, of arithmetic) *in its original, strict form* was shown by Gödel to be unrealizable. However, Gödel himself thought that the program for establishing the consistency of arithmetic might be salvageable through an enlargement of the domain of objects admitted into finitistic metamathematics. That is, by allowing finite manipulations of suitably chosen *abstract* objects in addition to the concrete ones Gödel hoped to strengthen finitistic metamathematics sufficiently to enable the consistency of arithmetic to be demonstrable within it. In 1958 he achieved his goal, constructing a consistency proof for arithmetic within a finitistic, but not

---

[14]Hilbert actually asserted that "no one will ever be able to expel us from the paradise that Cantor has created for us." There is no question that "Cantor's Paradise" furnishes the ideal site on which to build Hilbert's hotel (see the previous chapter.)

[15]It should be emphasized that Hilbert was *not* claiming that (classical) mathematics *itself* was meaningless, only that the formal system representing it was to be so regarded.

[16]A sketch of Gödel's arguments is given in Appendix 2.

strictly concrete, metamathematical system admitting, in addition to concrete objects (numbers), abstract objects such as functions, functions of functions, etc., over finite objects. So, although Hilbert's program cannot be carried out in its original form, for arithmetic at least Gödel showed that it can be carried out in a weakened form by countenancing the use of suitably chosen abstract objects.

As for the doctrine of "formalism" itself, this was for Hilbert (who did not use the term, incidentally) not the claim that mathematics could be *identified* with formal axiomatic systems. On the contrary, he seems to have regarded the role of formal systems as being to provide distillations of mathematical practice of a sufficient degree of precision to enable their formal features to be brought into sharp focus. The fact that Gödel succeeded in showing that certain features (e.g. consistency) of these logical distillations could be *expressed*, but *not demonstrated* by finitistic means does not undermine the essential cogency of Hilbert's program.

Intuitionism.

A third tendency in the philosophy of mathematics to emerge in the twentiethcentury, *intuitionism*, is largely the creation of Brouwer, already mentioned in connection with his contibutions to topology. Like Kant, Brouwer believed that mathematical concepts are admissible only if they are adequately grounded in *intuition*, that mathematical theories are significant only if they concern entities which are constructed out of something given immediately in intuition, that mathematical definitions must always be constructive, and that the completed infinite is to be rejected. Thus, like Kant, Brouwer held that mathematical theorems are synthetic *a priori* truths. In *Intuitionism and Formalism* (1912), while admitting that the emergence of noneuclidean geometry had discredited Kant's view of space, he maintained, in opposition to the logicists (whom he called "formalists") that arithmetic, and so all mathematics, must derive from the *intuition of time*. In his own words:

> Neointuitionism considers the falling apart of moments of life into qualitatively different parts, to be reunited only while remaining separated by time, as the fundamental phenomenon of the human intellect, passing by abstracting from its emotional content into the fundamental phenomenon of mathematical thinking, the intuition of the bare two-oneness. This intuition of two-oneness, the basal intuition of mathematics, creates not only the numbers one and two, but also all finite ordinal numbers, inasmuch as one of the elements of the two-oneness may be thought of as a new two-oneness, which process may be repeated indefinitely; this gives rise still further to the smallest infinite ordinal ω . Finally this basal intuition of mathematics, in which the connected and the separate, the continuous and the discrete are united, gives rise immediately to the intuition of the linear continuum, i.e., of the "between", which is not exhaustible by the interposition of new units and which can therefore never be thought of as a mere collection of units. In this way the apriority of time does not only qualify the properties of arithmetic as synthetic a priori judgments, but it does the same for those of geometry, and not only for elementary two- and three-dimensional geometry, but for non-euclidean and n-dimensional geometries as well. For since Descartes we have learned to reduce all these geometries to arithmetic by means of coordinates.

For Brouwer, intuition meant essentially what it did to Kant, namely, the mind's apprehension of what it has itself constructed; on this view, the only acceptable mathematical proofs are *constructive*. A constructive proof may be thought of as a kind of "thought experiment" —the performance, that is, of an experiment in imagination. According to *Arend Heyting* (1898–1980), a leading member of the intuitionist school,

> Intuitionistic mathematics consists ... in mental constructions; a mathematical theorem expresses a purely empirical fact, namely, the success of a certain construction. "$2 + 2 = 3 + 1$" must be read as an abbreviation for the statement "I have effected the mental construction indicated by '$2 + 2$' and '$3 + 1$' and I have found that they lead to the same result."

From passages such as these one might infer that for intuitionists mathematics is a purely subjective activity, a kind of introspective reportage, and that each mathematician has a personal mathematics. Certainly they reject the idea that mathematical thought is dependent on any special sort of language, even, occasionally, claiming that, at bottom, mathematics is a "languageless activity". Nevertheless, the fact that intuitionists evidently regard mathematical theorems as being valid for all intelligent beings indicates that for them mathematics has, if not an objective character, then at least a *transsubjective* one.

The major impact of the intuitionists' program of constructive proof has been in the realm of *logic*. Brouwer maintained, in fact, that the applicability of traditional logic to mathematics

> was caused historically by the fact that, first, classical logic was abstracted from the mathematics of the subsets of a definite finite set, that, secondly, an a priori existence independent of mathematics was ascribed to the logic, and that, finally, on the basis of this supposed apriority it was unjustifiably applied to the mathematics of infinite sets.

Thus Brouwer held that much of modern mathematics is based, not on sound reasoning, but on an illicit extension of procedures valid only in the restricted domain of the finite. He therefore embarked on the heroic course of setting the whole of existing mathematics aside and starting afresh, using only concepts and modes of inference that could be given clear intuitive justification. He hoped that, once enough of the program had been carried out, one could discern the logical laws that intuitive, or constructive, mathematical reasoning actually obeys, and so be able to compare the resulting *intuitionistic, or constructive, logic*[17] with classical logic.

The most important features of constructive mathematical reasoning are that *an existential statement can be considered affirmed only when an instance is produced*,[18] and—as a consequence—*a disjunction can be considered affirmed only when an explicit one of the disjuncts is demonstrated*. A striking consequence of this is that, as far as properties of (potentially) infinite domains are concerned, *neither the classical law of excluded middle*[19] *nor the law of strong reductio ad absurdum*[20] *can be*

---

[17]This is not to say that Brouwer was primarily interested in *logic*, far from it: indeed, his distaste for formalization led him not to take very seriously subsequent codifications of intuitionistic logic.

[18] Hermann Weyl said of nonconstructive existence proofs that "they inform the world that a treasure exists without disclosing its location."

[19]This is the assertion that, for any proposition p, either p or its negation $\neg p$ holds.

[20]This is the assertion that, for any proposition p, $\neg\neg p$ implies p.

accepted without qualification. To see this, consider for example the existential statement *there exists an odd perfect number* (i.e., an odd number equal to the sum of its proper divisors) which we shall write as $\exists n P(n)$. Its contradictory is the statement $\forall n \neg P(n)$. Classically, the law of excluded middle then allows us to affirm the disjunction

$$\exists n P(n) \vee \forall n \neg P(n) \tag{1}$$

Constructively, however, in order to affirm this disjunction we must *either* be in a position to affirm the first disjunct $\exists n P(n)$, i.e., to possess, or have the means of obtaining, an odd perfect number, *or* to affirm the second disjunct $\forall n \neg P(n)$, i.e. to possess a demonstration that no odd number is perfect. Since at the present time mathematicians have neither of these[21], the disjunction (1), and *a fortiori* the law of excluded middle is not constructively admissible.

It might be thought that, if in fact the second disjunct in (1) is *false*, that is, not every number falsifies P, then we can actually find a number satisfying P by the familiar procedure of testing successively each number 0, 1, 2, 3,... and breaking off when we find one that does: in other words, that from $\neg \forall n \neg P(n)$ we can infer $\exists n P(n)$. Classically, this is perfectly correct, because the *classical* meaning of $\neg \forall n \neg P(n)$ is "$P(n)$ will not as a matter of *fact* be found to fail for every number n." But *constructively* this latter statement has no meaning, because it presupposes that every natural number *has already been constructed* (and checked for whether it satisfies P). Constructively, the statement must be taken to mean something like "we can derive a contradiction from the supposition that we could prove that $P(n)$ failed for every n." From this, however, we clearly cannot extract a guarantee that, by testing each number in turn, we shall eventually find one that satisfies P. So we see that the law of strong reductio ad absurdum also fails to be constructively admissible.

As a simple example of a classical existence proof which fails to meet constructive standards, consider the assertion

there exists a pair of irrational real numbers a,b such that a^b is rational.

Classically, this can be proved as follows:

let $b = \sqrt{2}$; then b is irrational. If b^b is rational, let $a = b$; then we are through. If b^b is irrational, put $a = b^b$; then $a^b = 2$, which is rational.

But in this proof we have not *explicitly identified* a; we do not know, in fact, whether $a = 2$ or[22] $a = \sqrt{2}^{\sqrt{2}}$, and it is therefore constructively unacceptable.

Thus we see that constructive reasoning differs from its classical counterpart in that it attaches a stronger meaning to some of the logical operators. It has become

---

[21]And indeed may never have; as observed in Chapter 3, little if any progress has been made on the ancient problem of the existence of odd perfect numbers.

[22]In fact a much deeper argument shows that $2^{\sqrt{2}}$ is irrational, and is therefore the correct value of a.

customary, following Heyting, to explain this stronger meaning in terms of the primitive relation *a is a proof of p*, between mathematical constructions *a* and mathematical assertions *p*. To assert the *truth* of *p* is to assert that one has a construction *a* such that *a* is a proof of p^{23}. The meaning of the various logical operators in this scheme is spelt out by specifying how proofs of composite statements depend on proofs of their constituents. Thus, for example,

> *a* is a proof of $p \wedge q$ means: *a* is a pair (b, c) consisting of a proof *b* of *p* and *c* of *q*;
>
> *a* is a proof of $p \vee q$ means: *a* is a pair (b, c) consisting of a natural number *b* and a construction *c* such that, if $b = 0$, then *c* is a proof of *p*, and if $b \neq 0$, then *c* is a proof of *q*;
>
> *a* is a proof of $p \rightarrow q$ means: *a* is a construction that converts any proof of *p* into a proof of *q*;
>
> *a* is a proof of $\neg p$ means: *a* is a construction that shows that no proof of *p* is possible.

It is readily seen that, for example, the law of excluded middle is not generally true under this ascription of meaning to the logical operators. For a proof of $p \vee \neg p$ is a pair (b,c) in which *c* is either a proof of *p* or a construction showing that no proof of *p* is possible, and there is nothing inherent in the concept of mathematical construction that guarantees, for an arbitrary proposition *p*, that either will ever be produced.

As shown by Gödel in the 1930s, it is possible to represent the strengthened meaning of the constructive logical operators in a classical system augmented by the concept of *provability*. If we write $\Box p$ for "*p is provable*", then the scheme below correlates constructive statements with their classical translates.

Constructive	*Classical*
$\neg p$	$\Box \neg \Box p$
$p \wedge q$	$\Box p \wedge \Box q$
$p \vee q$	$\Box p \vee \Box q$
$p \rightarrow q$	$\Box(\Box p \rightarrow \Box q)$

The translate of the sentence $p \vee \neg p$ is then $\Box p \vee \Box\Box\neg\Box p$, which is (assuming $\Box\Box p \leftrightarrow \Box p$) equivalent to $\neg\Box p \rightarrow \Box\neg\Box p$, that is, to the assertion

> *if p is not provable, then it is provable that p is not provable.*

---

[23] Here by *proof* we are to understand a mathematical construction that establishes the assertion in question, *not* a derivation in some formal system. For example, a proof of $2 + 3 = 5$ in this sense consists of successive constructions of 2, 3 and 5, followed by a construction that adds 2 and 3, finishing up with a construction that compares the result of this addition with 5.

The fact that there is no *a priori* reason to accept this "solubility" principle lends further support to the intuitionists' rejection of the law of excluded middle.

Another interpretation of constructive reasoning is provided by *Kolmogorov's calculus of problems* (*A. N. Kolmogorov*, 1903–1987) If we denote problems by letters and $a \wedge b$, $a \vee b$, $a \to b$, $\neg a$ are construed respectively as the problems

> *to solve both a and b*
> *to solve at least one of a and b*
> *to solve b, given a solution of a*
> *to deduce a contradiction from the hypothesis that a is solved,*

then a formal calculus can be set up which coincides with the constructive logic of propositions.

Although intuitionism in Brouwer's original sense has not been widely adopted as a philosophy of mathematics, the constructive viewpoint associated with it has been very influential[24]. The intuitionistic logical calculus has also come under intensive investigation. If we compare the law of excluded middle with Euclid's fifth postulate, then intuitionistic logic may be compared with *neutral* geometry—geometry, that is, without the fifth postulate—and classical logic to Euclidean geometry. Just as noneuclidean geometry revealed a "strange new universe", so intuitionistic logic has allowed new features of the logico-mathematical landscape—invisible through the lens of classical logic—to be discerned. Intuitionistic logic has proved to be a subtle instrument, more delicate than classical logic, for investigating the mathematical world.[25]

*

Despite the fact that Logicism, Intuitionism and Formalism cannot be held to provide complete accounts of the nature of mathematics, each gives expression to an important partial truth about that nature: *Logicism,* that mathematical truth and logical demonstration go hand in hand; *Intuitionism,* that mathematics originates in the performance of mental constructions secured by intuitive evidence; and finally *Formalism,* that the results of these constructions are presented through the medium of formal symbols.

---

[24] Remarkably, it is also the logic of *smooth infinitesimal analysis*: see Appendix 3.

[25] In a famous remark opposing intuitionism, Hilbert said "to deny the mathematician the use of the law of excluded middle would be to deny the boxer the use of his fists." But with experience in using the refined apparatus of intuitionistic logic one comes to regard Hilbert's simile as inappropriate. It would, perhaps, be more apposite to compare the mathematician's frustration in being denied the use of the law of excluded middle with the frustration a nineteenth century surgeon might well have felt when denied the use of a butcher knife, or a twentieth century general the use of nuclear weapons.

APPENDIX 1

THE INSOLUBILITY OF SOME GEOMETRIC
CONSTRUCTION PROBLEMS

In this appendix we show that the problems of doubling the cube, trisecting an arbitrary angle, and producing the side of a regular heptagon cannot be solved using Euclidean tools.

We begin by introducing the notion of a *constructible* real number. Start with two points in the Cartesian plane at unit distance apart: for convenience we may take these to be the points $(0, 0)$ and $(1, 0)$ on the x-axis. A real number α is said to be *constructible* if the point $(\alpha, 0)$ is obtainable from the given points $(0, 0)$ and $(1, 0)$ by means of a finite number of applications of straightedge and compasses, subject to the Euclidean rules (see Chapter 2 for these). It is easily shown—and we leave this as an exercise to the reader—that if α and β are constructible numbers, so are $\alpha + \beta$, $\alpha - \beta$, $\alpha\beta$, and, if $\beta \neq 0$, α/β. Accordingly the set of constructible numbers forms a *field*.

We now prove the

Lemma. Suppose that F is a field of real numbers. Let k be a positive member of F such that $\sqrt{k}$ is not in F. Then the set $F(\sqrt{k})$ consisting of all numbers of the form $\alpha + \beta\sqrt{k}$ with α, β in F is also a field, called a *quadratic extension* of F.

Proof. It is clearly enough to show that, if $x \in F(\sqrt{k})$ and $x \neq 0$, then $1/x \in F(\sqrt{k})$. To this end suppose that $x = \alpha + \beta\sqrt{k}$, with α, $\beta \in F$. If $x \neq 0$, then also $\alpha \neq 0$ or $\beta \neq 0$, and so $\alpha^2 - \beta^2 k \neq 0$. Hence

$$\frac{1}{x} = \frac{1}{\alpha + \beta\sqrt{k}} = \frac{\alpha - \beta\sqrt{k}}{(\alpha + \beta\sqrt{k})(\alpha - \beta\sqrt{k})} = \frac{\alpha}{\alpha^2 - \beta^2 k} - \frac{\beta\sqrt{k}}{\alpha^2 - \beta^2 k} \, .$$

and this last quantity is clearly a member of F. This proves the lemma.

A *quadratic surd* is a real number which can be obtained in a finite number of steps from the numbers 0 and 1 using only the operations $+, \times, -, \div$ and the extraction of square roots. It is clear that the quadratic surds form a field. Moreover, it is quickly seen that a given real number a is a quadratic surd if and only if we can find a finite sequence of fields $F_0, F_1, ..., F_n$, which we call a *formation sequence* for a, such that $a \in F_n$, F_0 is the field $\mathbf{Q}$ of rational numbers, and each F_i is a quadratic extension of F_{i-1}. For example, if

$$a = \sqrt{2} + \sqrt{3} + \sqrt{5}$$

then F_0, F_1, F_2, F_3, F_4 *is* a formation sequence for a, where

$$F_0 = \mathbf{Q}, F_1 = \mathbf{Q}(\sqrt{2}), F_2 = F_1(\sqrt{3}), F_3 = F_2(\sqrt{5}), F_4 = F_3(\sqrt{2} + \sqrt{3} + \sqrt{5}).$$

We can now prove the crucial

Theorem. The constructible numbers are precisely the quadratic surds.

Proof. The field $\mathbf{Q}$ of rational numbers obviously consists of constructible numbers. Moreover, if k is constructible, so is $\sqrt{k}$, as can be seen from the diagram below:

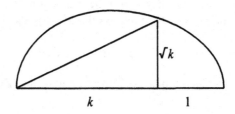

It follows that, if F is a field of constructible numbers, then each quadratic extension of F also consists of constructible numbers. Now let a be a quadratic surd, and let $F_0, ..., F_n$ be a formation sequence for a. Then $F_0 = \mathbf{Q}$ consists of constructible numbers, and so therefore do $F_1, ..., F_n$. Thus a, as a member of F_n, must be constructible.

To prove the converse, we observe that, if F is any field of real numbers, the points of intersection of lines and circles of F have coordinates which are members of some quadratic extension of F. (Here a *line of F* is any line passing through two points whose coordinates are in F, and a *circle of F* any circle whose centre has coordinates in F, and the length of whose radius is in F.) To prove this, notice that the coordinates (x, y) of such points of intersection are obtained by solving two simultaneous equations, each of which has one of the forms

$$\alpha x + \beta y + \gamma = 0 \qquad x^2 + y^2 + \varepsilon x + \eta y + \zeta = 0,$$

where $\alpha, \beta, \gamma, \varepsilon, \eta, \zeta$ are all members of F. By solving these equations explicitly for x and y we see immediately that they will both be members of some quadratic extension of F. Thus, starting with the points $(0, 0)$ and $(1, 0)$, straightedge and compass constructions can lead only to points whose coordinates lie in members of some sequence $F_0, F_1, ..., F_n$, with $F_0 = \mathbf{Q}$ and each F_i a quadratic extension of F_{i-1}. Accordingly each constructible number is a quadratic surd, and the theorem is proved.

Now we can show that *doubling the cube* cannot be performed with Euclidean tools. Taking the given cube to have side of unit length, the cube with double the volume will have side $\sqrt[3]{2}$. Thus we must show that this number is not constructible; by the theorem above this is tantamount to showing that it is not a quadratic surd. In fact we shall prove the ostensibly[1] stronger result that *no solution to the equation $x^3 - 2 = 0$ can be a quadratic surd.*

For suppose that a solution x to the equation $x^3 - 2 = 0$ were a quadratic surd, and let $F_0, \ldots, F_n$ be a formation sequence for x. Since (as is easily shown) $\sqrt[3]{2}$ is irrational, it cannot be a member of $F_0 = \mathbf{Q}$, and so n must be positive. Now assume that n is as small as it can be; thus x is in F_n but not in F_{n-1}. Now there is a member w of F_{n-1} such that $\sqrt{w}$ is not in F_{n-1} and $F_n = F_{n-1}(\sqrt{w})$, so that

$$x = p + q\sqrt{w},$$

for some $p, q \in F_{n-1}$.

We next show that, if $x = p + q\sqrt{w}$ is a solution of $x^3 - 2 = 0$, then so is $y = p - q\sqrt{w}$. For since x is in F_n, it follows that $x^3 - 2$ is also in F_n, and so

$$x^3 - 2 = a + b\sqrt{w}, \tag{1}$$

with $a, b \in F_{n-1}$. By substituting for x, and recalling that $\sqrt{w}$ is not in F_{n-1}, we easily find

$$a = p^3 + 3p^2qw - 2,$$
$$b = 3p^2q + q^3w.$$

Now put $y = p - q\sqrt{w}$. Then a substitution of $-q$ for q in the expressions for a and b above gives

$$y^3 - 2 = a - b\sqrt{w}.$$

Now x was assumed to satisfy $x^3 - 2 = 0$, so, by (1), we must have $a + b\sqrt{w} = 0$. But since $\sqrt{w} \notin F_{n-1}$, it follows that $a = b = 0$. (For if $b \neq 0$, then $\sqrt{w} = a/b$ which is a member of F_{n-1}. So $b = 0$ and $a = -b\sqrt{w} = 0$.) Therefore

$$y^3 - 2 = a - b\sqrt{w} = 0,$$

so that y is also a solution to $x^3 - 2 = 0$, as claimed.

Since $q \neq 0$ (for otherwise $x = p + q\sqrt{w}$ would be in F_{n-1}), it follows that x and y are distinct, so that the equation $x^3 - 2 = 0$ would have *two* real roots. But this contradicts the fact, evident from graphical considerations, that this equation has *only one* real root. This contradiction shows that our original assumption that the equation

---

[1] Only ostensibly stronger since it is easily seen that the equation $x^3 - 2 = 0$ has only the one real root $\sqrt[3]{2}$; the other roots, being complex numbers, by definition cannot be quadratic surds.

has a solution which is a quadratic surd was incorrect, and we conclude from this that the doubling of the cube cannot be carried out with Euclidean tools.

We now turn to the problem of *trisecting an arbitrary angle*. Here it is simplest to regard an angle as given by its *cosine*: $g = \cos \theta$. Thus the problem of trisecting θ is equivalent to finding the quantity $x = \cos(\tfrac{1}{3}\theta)$. From trigonometry one knows that

$$\cos \theta = g = 4\cos^3(\tfrac{1}{3}\theta) - 3\cos(\tfrac{1}{3}\theta),$$

and so the problem of trisecting an angle θ given by $\cos \theta = g$ with Euclidean tools amounts to constructing a solution to the cubic equation

$$4z^3 - 3z - g = 0.$$

To show that this cannot be done in general, we take θ to be $60°$, so that $g = \tfrac{1}{2}$. The above equation then becomes

$$8z^3 - 6z - 1 = 0. \tag{2}$$

We show that this equation has no solution which is a quadratic surd. To do this we need a general

Lemma. If a cubic equation with rational coefficients has no rational root, it has no root which is a quadratic surd.

Proof. Let the given equation be

$$z^3 + az^2 + bz + c = 0,$$

with a, b, c rational. It is a well known—and easily established—algebraic fact that the roots x_1, x_2, x_3 of such an equation themselves satisfy the relation

$$x_1 + x_2 + x_3 = -a. \tag{3}$$

Now suppose that the given cubic equation has no rational root, but does have at least one root which is a quadratic surd. Such a root will have a formation sequence $F_0, ..., F_r$. Let n be the least length of a formation sequence in which a quadratic surd root x to the equation can be found in its last member F_n; then n must be positive since the equation has no rational roots (recall that $F_0 = \mathbf{Q}$). The root x can be written in the form

$$x = p + q\sqrt{w},$$

ith p, q, w, but not $\sqrt{w}$, in F_{n-1}. As before, one shows that

$$y = p - q\sqrt{w}$$

is also a root. But then, by (3), the third root u satisfies

$$u = -a - x - y.$$

Since $x + y = 2p$, it follows that

$$u = -a - 2p.$$

Since $-a - 2p \in F_{n-1}$, so also, therefore, is u. Thus the root u lies in the last member F_{n-1} of a formation sequence of length $n - 1$, contradicting the choice of n as the least such number. This contradiction shows that our supposition above was mistaken, and the Lemma is proved.

Next, we show that equation (2) has no quadratic surd root. By the Lemma, this reduces to showing that it has no rational roots. To this end, put $v = 2z$. Then (2) becomes

$$v^3 - 3v = 1. \tag{4}$$

Suppose that (4) had a rational solution of the form $v = r/s$, where we may assume that r and s have no common factors. Then

$$r^3 - 3s^2 r = s^3.$$

Therefore

$$s^3 = r(r^2 - 3s^2)$$

is divisible by r, which means that r and s have a common factor, contrary to assumption, unless $r = \pm 1$. Likewise, s^2 is a factor of

$$r^3 = s^2(s + 3r),$$

which means that r and s again have a common factor unless $s = \pm 1$. Thus 1 and -1 are the only rational numbers which could satisfy the equation. But it is clear that neither of these satisfies it, so that no rational satisfies it, and we are done.

Finally, we turn to the problem of constructing the side of a regular heptagon, which we may take as being inscribed in the unit circle in the complex plane. If each vertex has coordinates x, y, then we know that $z = x + iy$ is a root of the equation $z^7 - 1 = 0$. One root of this equation is $z = 1$, and the others are the roots of the equation

$$(z^7 - 1)/(z - 1) = z^6 + z^5 + z^4 + z^3 + z^2 + z + 1 = 0.$$

Dividing this by z^3, we obtain the equation

$$z^3 + 1/z^3 + z^2 + 1/z^2 + z + 1/z + 1 = 0.$$

This may be written in the form

$$(z + 1/z)^3 + (z + 1/z)^2 - 2(z + 1/z) - 1 = 0.$$

Writing y for $z + 1/z$, this last equation becomes

$$y^3 + y^2 - 2y - 1 = 0. \tag{5}$$

Now we have seen in Chapter 3 that z, the seventh root of unity, is given by $z = \cos\theta + i\sin\theta$, where $\theta = 360°/7$. We also know that $1/z = \cos\theta - i\sin\theta$, so that $y = z + 1/z = 2\cos\theta$. It follows that the constructibility of y is equivalent to that of $\cos\theta$. Accordingly, if we can show that y is not constructible, we will also have shown that z, and so also the side of the heptagon, is not constructible. By the theorem and the second lemma above, to do this we need merely show that equation (5) has no rational roots. So suppose that r/s were a rational root of (5), with r and s possessing no common factor. Substitution into (5) then gives

$$r^3 + r^2s - 2rs^2 - s^3 = 0,$$

and, as above, it follows that r^3 is divisible by s, and s^3 by r. Since r and s have no common factor, each must be ± 1, so that $y = \pm 1$ likewise. But neither of these values of y satisfies (5). Therefore (5) has no rational roots and we are finished.

APPENDIX 2

THE GÖDEL INCOMPLETENESS THEOREMS

We sketch proofs of Gödel's theorems, obtaining along the way an important result of Tarski on the undefinability of mathematical truth.

We begin by setting up a *formal language* in which arithmetical statements can be written down in a precise way. This language will be denoted by the symbol **L** and called the *language of arithmetic*. As is the case with any language, we must first specify its *alphabet*: that of **L** consists of the following *symbols*:

Arithmetical variables: x_1, x_2, x_3, ...
Arithmetical constants: **0, 1, 2,** ...
Arithmetical operation symbols: +, ×
Equality symbol: =
Logical operators: ∧ *(conjunction, "and")*
      ∨ *(disjunction, "or")*
      ¬ *(negation, "not"),*
      → *(implication, "if... then")*
      ↔ *(bi-implication, "is equivalent to")*
      ∀ *(universal quantifier, "for all")*
      ∃ *(existential quantifier, "there exists")*
Punctuation symbols: (,) , [,], commas, etc.

Expressions of **L**—*arithmetical expressions*—are built up by stringing together finite sequences of symbols. As in any language, only certain expressions of **L** will be deemed meaningful or *well-formed*. These are the *terms*, the arithmetical counterparts of nouns, and the *formulas*, the arithmetical counterparts of declarative assertions.

Arithmetical terms, or simply *terms*, are specified by means of the following rules:

(i) Any arithmetical variable or constant standing by itself is a term.

(ii) If t and u are terms, so too are the expressions $t + u$, $t \times u$.

(iii) An expression is a term when, and only when, it follows that it is one from finitely many applications of clauses (i) and (ii).

Thus, for example, each of the expressions 6, $x_1 + 2$ and $(x_1 \times x_5) + 27$ is a term.

Arithmetical formulas, or simply *formulas*, are specified by means of the following rules:

(a) For any terms t, u, the expression $t = u$ is a formula.

(b) If A and B are formulas, so too are all the expressions $\neg A$, $A \wedge B$, $A \vee B$, $A \to B$, $A \leftrightarrow B$, $\forall x_n A$, $\exists x_n A$, for any numerical variable x_n.

(c) An expression is a formula when, and only when, it follows that it is one from finitely many applications of clauses (a) and (b).

Thus, for instance, each of the following expressions is a formula: $x_1 + 2 = 3$, $\exists x_1 (x_1 + 2 = 3)$, $\forall x_1 \forall x_5 (x_1 \times x_5 = 3)$.

Terms assume *numerical values* and formulas *truth values* (truth or falsehood) when their constituent symbols are interpreted in the natural way. In this natural interpretation, each arithmetical variable is assigned some arbitrary but fixed integer as value, and then each arithmetical constant n is interpreted as the corresponding integer n, the arithmetical operation symbols $+$ and $\times$ as addition and multiplication of integers, respectively, the equality symbol as identity of integers, and finally the logical operators as the corresponding logical particles of ordinary discourse. For example, if we assign the values 3 to x_1 and 7 to x_5, then the resulting values assigned to the three terms above are, respectively, $3 + 2$, i.e., 5; 6; and $(3 \times 7) + 27$, i.e., 48. Under the same assignment of values, the resulting truth values assigned to the three formulas above are, respectively,

THE TRUTH VALUE OF THE STATEMENT $3 + 2 = 3$, I.E., *FALSEHOOD;*
THE TRUTH VALUE OF THE STATEMENT *THERE IS A NUMBER WHOSE SUM WITH 2 EQUALS 3*,
I.E., *TRUTH* ;
THE TRUTH VALUE OF THE STATEMENT *THE PRODUCT OF ANY PAIR OF NUMBERS EQUALS 3*,
I.E., *FALSEHOOD.*

In place of the clumsy locution "the truth value of A is truth (or falsehood)" we shall usually employ the phrase " A is *true* (or *false*)." Note that then a formula A is true exactly when its negation $\neg A$ is false.

We observe that, while the truth or falsehood of the first of these formulas is dependent on the values assigned to the variables occurring in it (in this case, just x_1), the truth values of the second two are *independent* of the values assigned to such variables. Formulas having this independence property are called *sentences*: they may be regarded as making simple declarative assertions—either true or false—about the system of natural numbers. Formally speaking, a sentence is a formula in which each occurrence of a variable x is accompanied by the occurrence of a corresponding "quantifier" expression of the form $\forall x$ or $\exists x$.

Occurrences of variables in formulas not accompanied by a corresponding quantifier expression are called *free* occurrences: for example, the occurrence of th variable x_1 in the first formula above is free, but those in the second and third formul are not. We write $A(x_1 \ldots, x_n)$ for any formula A in which at most the variables $x_1, \ldots, x_n$ have free occurrences, and, for any natural numbers $m_1, \ldots, m_n$, we write $A(m_1, \ldots, m_n)$ for the formula (evidently a sentence) obtained by substituting m_1 for $x_1, \ldots, m_n$ for x_n at each of the latter's free occurrences in A. Thus, for example, if $A(x_1, x_2, x_3, x_4)$ is the formula

$$\exists x_1 (x_1 + x_2 = 4 \wedge x_3 \times x_4 = 7),$$

then $A(1, 5, 7, 8)$ is the sentence

$$\exists x_1 (x_1 + 5 = 4 \wedge 7 \times 8 = 7).$$

We shall employ similar notational conventions for terms.

We next assign *code numbers* to the symbols and expressions of the language of arithmetic in the following way. Suppose that its symbols, excluding variables and constants, are k in number. To these symbols we assign, in some initially arbitrary but subsequently fixed manner, the code numbers $0, 1, \ldots, k - 1$. Then to each numerical variable x_n we assign the code number $k + 2n$ and to each numerical constant n the code number $k + 2n + 1$. In this way each symbol s is assigned a code number which we shall denote by s^*. Finally, each expression $s_1 s_2 \ldots s_n$ is assigned the code number

$$2^{s^*} \times 3^{s^*} \times \ldots \times p_n^{s^*},$$

where p_n is the nth prime number. In this way each expression is assigned a unique positive integer as its code, and, conversely, every positive integer is the code of some unique expression.

We shall use the symbol A_n to denote the arithmetical expression with code number n.

Let P be a property of natural numbers: we write $P(m)$ to indicate that the number m has the property P. Similarly, if R is a relation among natural numbers, we write $R(m_1, \ldots, m_n)$ to indicate that the numbers $m_1, \ldots, m_n$ stand in the relation R. We shall often use the term "relation" to cover properties as well.

A relation R among natural numbers is called *arithmetically definable* if there is an arithmetical formula $A(x_1, \ldots, x_n)$ such that, for all numbers $m_1, \ldots, m_n$, we have

$$R(m_1, \ldots, m_n) \text{ iff}^{[1]} \text{ the sentence } A(m_1, \ldots, m_n) \text{ is true.}$$

In this case we say that the relation R *is defined by* the formula A. We extend this concept to arithmetical *expressions* by saying that a property of (or a relation among) such expressions is arithmetically *definable* if the corresponding property of (or relation among) their *code numbers* is so definable.

Now it can be shown without much difficulty that the property of being (the code number of) an arithmetical *formula*, or a *sentence*, is arithmetically definable. But

---

[1] We use "iff" as an abbreviation for the phrase "if and only if", that is, "is equivalent to".

what about the property of being a *true* sentence? We shall establish the remarkable result that this property is *not* arithmetically definable.

Since the assignment of code numbers to arithmetical expressions is evidently a wholly mechanical process, it is possible to compute, for any given formula[2] $A_m(x_1)$ with code number m, and any number n, the code number of the sentence $A_m(n)$. This computation is in turn arithmetically representable in the sense that one can construct an arithmetical term[3] $s(x_1, x_2)$ with the property that, for any numbers m, n, p,

the sentence $s(m, n) = p$ is true iff p is the code number of the sentence $A_m(n)$.

Now let S any collection of arithmetical sentences. We proceed to prove the

Arithmetical Truth Theorem. Suppose that S satisfies the following conditions:

(i) Each member of S is true.
(ii) The property of being (the code number of) a member of S is arithmetically definable.

Then there is a *true* sentence G of **L** such that neither G nor its negation $\neg G$ are members of S.

Proof. By assumption (ii) there is a formula $T(x_1)$ of **L** such that, for all numbers n,

$T(n)$ is true iff n is the code number of a sentence in S

iff A_n is in S.

Write $B(x_1)$ for the formula $\neg T(s(x_1, x_1))$ (i.e., the result of substituting $s(x_1, x_1)$ for all free occurrences of x_1 in $\neg T(x_1)$) and suppose that B has code number m. Then

B is the sentence A_m.

Next, let p be the natural number such that

$$p = s(m, m)$$

is a true sentence. Then, *by definition,* p is the code number of the sentence $A_m(m)$.

Now write G for $A_m(m)$. Then p is the code number of G, or, in other words,

---

[2]Recall that A_m stands for the formula with code number m. In writing $A_m(x_1)$ we are, accordingly, assuming that A_m has free occurrences of at most the variable x_1.

[3]Here "s" stands for "substitution."

$$G \text{ is } A_p.$$

Thus we have

$$G \text{ is true iff } A_m(m).\text{is true}$$

$$\text{iff } B(m) \text{ is true}$$

$$\text{iff } \neg T(s(m, m)) \text{ is true}$$

$$\text{iff } T(p) \text{ is false}$$

$$\text{iff } A_p \text{ is not in } S$$

$$\text{iff } G \text{ is not in } S.$$

We see from this G asserts of itself that it is not in S. It follows that G is true, for, if it were false, it would follow from the above that it was in S, and hence true by assumption (i). Since G is now true, again by the above it cannot be a member of S. Finally, since $\neg G$ must now be false, it cannot be a member of S since by assumption every member of the latter is true. The proof is complete.

By taking S in this theorem to be the collection of *all true arithmetical sentences*, we immediately obtain

Tarski's Theorem on the Undefinability of Truth. The property of being a true arithmetical sentence is not arithmetically definable.

We observe that the relevant sentence G in Tarski's theorem asserts "I am not in the set of true sentences", i.e., "I am false". Thus, like the sentence in the Liar paradox, G asserts its own falsehood.

Next, one can formulate the notion of a *proof* from a set of arithmetical sentences S and that of a formula *provable* from S in such a way that:

(1) if each member of S is true, so is each sentence provable from S;

(2) if the property of being a member of S is arithmetically definable, so is the property of being a sentence provable from S.

Then from the Arithmetical Truth Theorem one infers

Gödel's First Incompleteness Theorem (weak form). Let S be a set of true arithmetical sentences and suppose that the property of being a member of S is arithmetically definable. Then S is *incomplete*, i.e. there is a (true) arithmetical sentence G such that neither it nor its negation are provable from S.

To prove this we define S^* to be the set of all sentences provable from S. Then by (1) and (2) above, S^* consists of true sentences and the property of being a member of S^* is arithmetically definable. Accordingly we may apply the Arithmetical Truth Theorem to S^*: this yields a (true) sentence G such that neither G nor $\neg G$ are members of S^*, in other words, neither are provable from S.

The sentence G here will be seen to assert, not its own falsehood, but its own *unprovability* from S.

The import of this theorem may be stated in the following way. Suppose we think of our set S as a possible set of *axioms* for arithmetic, from which one might hope to be able to infer (at least in principle) all arithmetical truths. Then the theorem shows that one will *never* be able to construct within any language like L a set S of axioms for arithmetic which is *sound,* in the sense that each of its members is true, arithmetically *definable,* so that we know what to put in it, and *complete,* so that it can be used to prove or refute[4] any arithmetical sentence. It is possible for S to possess *any two* of these properties, *but not all three at once.*

By refining this argument one can significantly strengthen its conclusion. Let us call S *consistent* if no formula of the form $A \wedge \neg A$ (a *contradiction*) is provable from S. Let R be a relation defined by a formula A. We say that R is *S-definite* if, for any natural numbers $m_1, ..., m_n$ we have

$$R(m_1, ..., m_n) \text{ iff } A(m_1, ..., m_n) \text{ is provable from } S$$
$$\text{not } R(m_1, ..., m_n) \text{ iff } \neg A(m_1, ..., m_n) \text{ is provable from } S.$$

Let Q be the *substitution relation* among natural numbers, that is, the relation which obtains among those triples m_1, m_2, m_3 of numbers for which the sentence $s(m_1, m_2) = m_3$ is true. Then one can prove

Gödel's First Incompleteness Theorem (strong form). Suppose that S is consistent, the property of being a member of S is arithmetically definable, and the property Q and the property of being a formula provable from S are both S-definite. Then S is incomplete.

We sketch the proof of this theorem. As before, we take G to be an arithmetical sentence which asserts its own unprovability from S. For any formula A, let us write $S \vdash A$ for "A is provable from S".

Suppose that $S \vdash G$. Then because provability from S is S-definite, it follows that

$$S \vdash \text{"}G \text{ is provable from } S\text{"}.$$

---

[4]We say that a sentence is *refutable* if its negation is provable.

But the assertion "G is provable from S" is essentially just $\neg G$ (since G is essentially "G is unprovable from S"), so we get

$$S \vdash \neg G,$$

contradicting the supposed consistency of S. Therefore not $S \vdash G$.

Now suppose that $S \vdash \neg G$. Then since S is consistent, it follows that not $S \vdash G$ and because provability from S is S-definite, we get

$$S \vdash \text{"}G\text{ is unprovable from }S\text{"}.$$

Noting again that G is essentially the assertion "G is unprovable from S", it then follows that

$$S \vdash G,$$

contradicting the consistency of S. Hence also not $S \vdash \neg G$, and we are done.

The advantage of this strong form of the incompleteness theorem is that in it the "external" requirement that all the members of S be *true* has been replaced by the much weaker "internal" requirement that S be merely *consistent*. We may sum it up by saying that *any consistent definable set of axioms for arithmetic must be incomplete*.

In sketching the proof of this last theorem we established the implication

$$\textit{if } S \textit{ is consistent, then not } S \vdash G. \tag{*}$$

Now the assertion "S is consistent" can be expressed as an arithmetical sentence Con_S in the following way. Let n_0 be the code number of some demonstrably false sentence, $0 = 1$, say, and write $P(x_1)$ for the arithmetical formula defining the property of being the code number of a sentence provable from S. Then Con_S may be taken to be the sentence

$$\neg P(n_0).$$

It turns out that the proof of the implication (*) can be written down formally in the language of arithmetic. Recalling yet again that G is essentially the assertion "not $S \vdash G$", this yields a proof from S of the arithmetical sentence

$$Con_S \rightarrow G,$$

Now suppose that

$$S \vdash Con_S.$$

Then since, as we have seen,

$$S \vdash Con_S \rightarrow G,$$

it would follow that

$$S \vdash G.$$

But, by the First Incompleteness Theorem, if S is consistent, then *not* $S \vdash G$. This is a contradiction, and we infer

Gödel's Second Incompleteness Theorem. Under the same conditions as the strong form of the First Incompleteness Theorem, the arithmetical sentence Con_S expressing the consistency of S is not provable from S.

In other words, the consistency of any arithmetically definable consistent system of axioms for arithmetic is not demonstrable in the system itself. Thus the consistency of arithmetic—assuming that it is indeed consistent—can only be demonstrated by appeal to procedures which *transcend arithmetic,* that is, in which the infinite figures in some essential way. This discovery dealt a shattering blow to Hilbert's program for establishing the consistency of mathematics by "finitistic" means.

APPENDIX 3

THE CALCULUS IN SMOOTH INFINITESIMAL ANALYSIS

In the usual development of the calculus, for any differentiable function f on the real line **R**, $y = f(x)$, it follows from Taylor's theorem that the increment $\delta y = f(x + \delta x) - f(x)$ in y attendant upon an increment δx in x is determined by an equation of the form

$$\delta y = f'(x)\delta x + A(\delta x)^2, \tag{1}$$

where $f'(x)$ is the derivative of $f(x)$ and A is a quantity whose value depends on both x and δx. Now if it were possible to take δx so *small* (but not demonstrably identical with 0) that $(\delta x)^2 = 0$ then (1) would assume the simple form

$$f(x + \delta x) - f(x) = \delta y = f'(x)\,\delta x. \tag{2}$$

We shall call a quantity having the property that its square is zero a *nilsquare infinitesimal* or simply an *infinitesimal*. In *smooth infinitesimal analysis* "enough" infinitesimals are present to ensure that equation (2) holds *nontrivially* for *arbitrary* functions $f: \mathbf{R} \to \mathbf{R}$. (Of course (2) holds trivially in standard mathematical analysis because there 0 is the sole infinitesimal in this sense.) The meaning of the term "nontrivial" here may be explicated in following way. If we replace δx by the letter ε standing for an arbitrary infinitesimal, (2) assumes the form

$$f(x + \varepsilon) - f(x) = \varepsilon f'(x). \tag{3}$$

Ideally, we want the validity of this equation to be independent of ε, that is, given x, for it to hold for *all* infinitesimal ε. In that case the derivative $f'(x)$ may be *defined* as the unique quantity H such that the equation

$$f(x + \varepsilon) - f(x) = \varepsilon H$$

holds for all infinitesimal ε.

Setting $x = 0$ in this equation, we get in particular

$$f(\varepsilon) = f(0) + H\varepsilon, \tag{4}$$

224

for all ε. *It is equation (4) that is taken as axiomatic in smooth infinitesimal analysis.* Let us write Δ for the set of infinitesimals, that is,

$$\Delta = \{x: x \in \mathbf{R} \wedge x^2 = 0\}.$$

Then it is postulated that, for any $f: \Delta \to \mathbf{R}$, there is a *unique* $H \in \mathbf{R}$ such that equation (4) holds for all ε. This says that the graph of f is a straight line passing through $(0, f(0))$ with slope H. Thus any function on Δ is what mathematicians term *affine*, and so this postulate is naturally termed the *principle of infinitesimal affineness*. It means that Δ *cannot be bent or broken*: it is subject only to *translations and rotations*—and yet is not (as it would have to be in ordinary analysis) identical with a point. Δ may be thought of as an entity possessing position and attitude, but lacking true extension.

If we think of a function $y = f(x)$ as defining a curve, then, for any a, the image under f of the "infinitesimal interval" $\Delta + a$ obtained by translating Δ to a is straight and coincides with the tangent to the curve at $x = a$ (see figure immediately below). In this sense each curve is "infinitesimally straight".

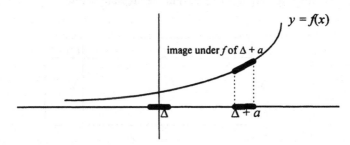

From the principle of infinitesimal affineness we deduce the important *principle of infinitesimal cancellation*, viz.

IF $\varepsilon a = \varepsilon b$ *FOR ALL* ε, THEN $a = b$.

For the premise asserts that the graph of the function $g: \Delta \to \mathbf{R}$ defined by $g(\varepsilon) = a\varepsilon$ has both slope a and slope b: the uniqueness condition in the principle of infinitesimal affineness then gives $a = b$. The principle of infinitesimal cancellation supplies the exact sense in which there are "enough" infinitesimals in smooth infinitesimal analysis.

From the principle of infinitesimal affineness it also follows that *all functions on* $\mathbf{R}$ *are continuous*, that is, *send neighbouring points to neighbouring points*. Here two points x, y on $\mathbf{R}$ are said to be neighbours if $x - y$ is in Δ, that is, if x and y differ by an infinitesimal. To see this, given $f: \mathbf{R} \to \mathbf{R}$ and neighbouring points x, y, note that $y = x + \varepsilon$ with ε in Δ, so that

$$f(y) - f(x) = f(x + \varepsilon) - f(x) = \varepsilon f'(x).$$

But clearly any multiple of an infinitesimal is also an infinitesimal, so $\varepsilon f'(x)$ is infinitesimal, and the result follows.

In fact, since equation (3) holds for any f, it also holds for its derivative f'; it follows that functions in smooth infinitesimal analysis are differentiable arbitrarily many times, thereby justifying the use of the term "smooth".

Let us derive a basic law of the differential calculus, the *product rule*:

$$(fg)' = f'g + fg'.$$

To do this we compute

$$(fg)(x + \varepsilon) = (fg)(x) + (fg)'(x) = f(x)g(x) + (fg)'(x),$$
$$(fg)(x + \varepsilon) = f(x + \varepsilon)g(x + \varepsilon) = [f(x) + f'(x)].[g(x) + g'(x)]$$
$$= f(x)g(x) + \varepsilon(f'g + fg') + \varepsilon^2 f'g'$$
$$= f(x)g(x) + \varepsilon(f'g + fg'),$$

since $\varepsilon^2 = 0$. Therefore $\varepsilon(fg)' = \varepsilon(f'g + fg')$, and the result follows by infinitesimal cancellation. This calculation is depicted in the diagram below.

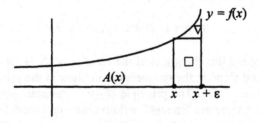

Next, we derive the *Fundamental Theorem of the Calculus*.

Let J be a closed interval $\{x: a \le x \le b\}$ in $\mathbf{R}$ and $f: J \to \mathbf{R}$; let $A(x)$ be the area under the curve $y = f(x)$ as indicated above. Then, using equation (3),

$$\varepsilon A'(x) = A(x + \varepsilon) - A(x) = \square + \nabla = \varepsilon f(x) + \nabla.$$

Now by infinitesimal affineness ∇ is a triangle of area $\frac{1}{2}\varepsilon.\varepsilon f'(x) = 0$. Hence $\varepsilon A'(x) = \varepsilon f(x)$, so that, by infinitesimal cancellation,

$$A'(x) = f(x).$$

A *stationary point* a in $\mathbf{R}$ of a function $f: \mathbf{R} \rightarrow \mathbf{R}$ is defined to be one in whose vicinity "infinitesimal variations" fail to change the value of f, that is, such that $f(a + \varepsilon) = f(a)$ for all ε. This means that $f(a) + \varepsilon f'(a) = f(a)$, so that $\varepsilon f'(a) = 0$ for all ε, whence it follows from infinitesimal cancellation that $f'(a) = 0$. This is *Fermat's rule*.

An important postulate concerning stationary points that we adopt in smooth infinitesimal analysis is the

Constancy Principle. If every point in an interval J is a stationary point of $f: J \rightarrow \mathbf{R}$ (that is, if f' is identically 0), then f is constant.

Put succinctly, "universal local constancy implies global constancy". It follows from this that two functions with identical derivatives differ by at most a constant.

In ordinary analysis the continuum $\mathbf{R}$ is connected in the sense that it cannot be split into two non empty subsets neither of which contains a limit point of the other. In smooth infinitesimal analysis it has the vastly stronger property of *indecomposability*: it cannot be split *in any way whatsoever* into two disjoint nonempty subsets. For suppose $\mathbf{R} = U \cup V$ with $U \cap V = \varnothing$. Define $f: \mathbf{R} \rightarrow \{0, 1\}$ by $f(x) = 1$ if $x \in U$, $f(x) = 0$ if $x \in V$. We claim that f is constant. For we have

$$(f(x) = 0 \text{ or } f(x) = 1) \quad \& \quad (f(x + \varepsilon) = 0 \text{ or } f(x + \varepsilon) = 1).$$

This gives 4 possibilities:

(i) $f(x) = 0$ & $f(x + \varepsilon) = 0$
(ii) $f(x) = 0$ & $f(x + \varepsilon) = 1$
(iii) $f(x) = 1$ & $f(x + \varepsilon) = 0$
(iv) $f(x) = 1$ & $f(x + \varepsilon) = 1$

Possibilities (ii) and (iii) may be ruled out because f is continuous. This leaves (i) and (iv), in either of which $f(x) = f(x + \varepsilon)$. So f is locally, and hence globally, constant, that is, constantly 1 or 0. In the first case $V = \varnothing$, and in the second $U = \varnothing$.

Partial derivatives can be defined in smooth infinitesimal analysis in a way similar to ordinary derivatives. For example, for arbitrary infinitesimals ε, η, we have the equations

$$f(x + \varepsilon, y) - f(x, y) = \varepsilon \, \partial f/\partial x, \qquad f(x, y + \eta) - f(x, y) = \eta \, \partial f/\partial y.$$

We use these in the derivation of the one-dimensional *heat equation*.

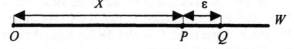

Thus suppose we are given a heated wire W. Let $T(x, t)$ be the temperature at a point P on it at time t. The heat content of the segment PQ is $k\epsilon T_{average}$, where k is a constant and $T_{average}$ is the average temperature along PQ. Now

$$T_{average} = \tfrac{1}{2}[T(x+\epsilon, t) + T(x, t)] = T(x, t) + \tfrac{1}{2}\epsilon\,(\partial T/\partial x)(x, t).$$

Therefore the heat content of PQ is

$$k\,[T(x, t) + \tfrac{1}{2}\epsilon\,(\partial T/\partial x)(x, t)] = k\,T(x, t).$$

So the change in the heat content of PQ between time t and time $t + \eta$ is given by

$$k\epsilon\,[T(x+\eta, t) - T(x, t)] = k\epsilon\eta\,(\partial T/\partial t)(x, t). \qquad (5)$$

Now the rate of flow of heat across P is, according to a basic law of heat conduction, proportional to the temperature gradient there, that is, equal to

$$m(\partial T/\partial x)(x, t),$$

where m is a constant. Similarly, the rate of flow of heat across Q is

$$m(\partial T/\partial x)(x + \epsilon, t),$$

So the heat transfer across P between the times t and $t + \eta$ is

$$m\eta\,(\partial T/\partial x)(x, t),$$

and across Q is

$$m\eta\,(\partial T/\partial x)(x + \epsilon, t).$$

So the net change of heat content in PQ between t and $t + \eta$ is

$$m\eta\,[(\partial T/\partial x)(x + \epsilon, t) - (\partial T/\partial x)(x, t)] = m\eta\epsilon\,(\partial/\partial x)(\partial T/\partial x) = m\eta\epsilon\,\partial^2 T/\partial x^2.$$

Equating this with (5), cancelling ϵ and η and writing $K = m/k$ yields the *one-dimensional heat equation*

$$\partial T/\partial t = K\,\partial^2 T/\partial x^2.$$

In conclusion, we observe that the postulates of smooth infinitesimal analysis are *incompatible with the law of excluded middle of classical logic* (*q.v.* Chapter 12). This incompatibility can be demonstrated in two ways, one informal and the other rigorous. First the informal argument. Consider the function f defined for real numbers x by $f(x) = 1$ if $x = 0$ and $f(x) = 0$ whenever $x \neq 0$. If the law of excluded middle held, each real number is then either equal or unequal to 0, so that the function f would be defined on the whole of **R**. But, considered as a function with domain **R**, f is clearly discontinuous. Since, as we know, in smooth infinitesimal analysis every function on **R** is continuous, f cannot have domain **R** there. So the law of excluded middle fails in smooth infinitesimal analysis. To put it succinctly, *universal continuity implies the failure of the law of excluded middle.*

Here now is the rigorous argument. We show that the failure of the law of excluded middle can be derived from the principle of infinitesimal cancellation. To begin with, if $x \neq 0$, then $x^2 \neq 0$, so that, if $x^2 = 0$, then necessarily not $x \neq 0$. This means that

$$\text{for all infinitesimal } \varepsilon, \text{ not } \varepsilon \neq 0. \tag{*}$$

Now suppose that the law of excluded middle were to hold. Then we would have, for any ε, either $\varepsilon = 0$ or $\varepsilon \neq 0$. But (*) allows us to eliminate the second alternative, and we infer that, for all ε, $\varepsilon = 0$. This may be written

$$\text{for all } \varepsilon, \ \varepsilon.1 = \varepsilon.0,$$

from which we derive by infinitesimal cancellation the falsehood $1 = 0$. So again the law of excluded middle must fail.

The "internal" logic of smooth infinitesimal analysis is accordingly not full classical logic. It is, instead, *intuitionistic* logic, that is, the logic—described in Chapter 12—derived from the constructive interpretation of mathematical assertions. In our brief sketch we did not notice this "change of logic" because, like much of elementary mathematics, the topics we discussed are naturally treated by constructive means such as direct computation.

APPENDIX 4

THE PHILOSOPHICAL THOUGHT OF A GREAT MATHEMATICIAN:

HERMANN WEYL

As we have observed, mathematics and philosophy are closely linked, and several great mathematicians who were at the same time great philosophers come to mind—Pythagoras, Descartes and Leibniz, for instance. One great mathematician of the modern era in whose thinking philosophy played a major role was *Hermann Weyl* (1885–1955), whose work encompassed analysis, number theory, topology, differential geometry, relativity theory, quantum mechanics, and mathematical logic. His many writings are informed by a vast erudition, an acute philosophical awareness, and even, on occasion, a certain playfulness. No matter what the subject may be—mathematics, physics, philosophy—Weyl's writing fascinates both by the depth of insight it reveals and by its startling departures from academic convention. Who else, for instance, would have the daring to liken (as he does in the discussion of Space and Time in his *Philosophy of Mathematics and Natural Science*), a coordinate system to "the residue of the annihilation of the ego"[1]? Or then (somewhat further on in the same discussion) to express the belief in the impossibility of a completely objective account of individual consciousness by the assertion "...it is shattered by Judas' desperate outcry, 'Why did *I* have to be Judas?'"[2].

In this final Appendix we trace Weyl's philosophical thinking on mathematics and science through quotations from his works. It is a fascinating and instructive journey.

Weyl's philosophical outlook was strongly influenced by that of *Edmund Husserl* (1859–1938), the creator of the philosophy known as *phenomenology*. The principal tenet of phenomenology is that the only things which are directly given to us, that we can know completely, are objects of consciousness. It is with these that philosophy, and all knowledge, must begin. This is acknowledged by Weyl in the introduction to *Space-Time-Matter* (1919), his famous book on the theory of relativity.

---

[1] Weyl, *Philosophy*, 123. This metaphor seems first to have appeared in *The Continuum*, where Weyl asserts on p. 94:

> The coordinate system is the unavoidable residue of the eradication of the ego in that geometrico-physical world which reason sifts from the given using "objectivity" as its standard—a final scanty token in this objective sphere that existence is only given and *can* only be given as the intentional content of the processes of consciousness of a pure, sense-giving ego.

[2] *Ibid.*, 124–125.

Modestly describing his remarks as "a few reflections of a philosophical character," he observes that the objective world "constructed" by mathematical physics cannot of necessity coincide with the subjective world of qualities given through perception:

> Expressed as a general principle, this means that the real world, and every one of its constituents, are, and can only be, given as intentional objects of acts of consciousness. The immediate data which I receive are the experiences of consciousness in just the form in which I receive them ... we may say that in a sensation an object, for example, is actually physically present for me—to whom that sensation relates—in a manner known to everyone, yet, since it is characteristic, it cannot be described more fully. (Weyl, *Space-Time-Matter*, 4)

His phenomenological orientation is proclaimed still more emphatically when he goes on to say:

> ...the datum of consciousness is the starting point at which we must place ourselves if we are to understand the absolute meaning of, as well as the right to, the supposition of reality ... "Pure consciousness" is the seat of what is philosophically a priori. (*ibid.*, 5)

Later, he asserts that

> Time is the ... form of the stream of consciousness ...and space... the form of external material reality. (*ibid.*, 5)

And in a memorable passage he describes how these two opposed facets of existence ineluctably come to be fused, so leading to the amalgamation of time and space, of which the theory of relativity is (thus far) the deepest expression:

> ... if the worlds of consciousness and of transcendental reality were totally different from one another, or, rather, if only the passive act of perception bridged the gulf between them, the state of affairs would remain as I have represented it, namely, on the one hand a consciousness rolling on in the form of a lasting present, yet spaceless; on the other, a reality spatially extended, yet timeless, of which the former contains but a varying appearance. Antecedent to all perception there is in us the experience of effort and opposition, of being active and being passive. For a person leading a natural life of activity, perception serves above all to place clearly before his consciousness the definite point of the action he wills, and the source of the opposition to it. As the doer and endurer of actions I become a single individual with a psychical reality attached to a body which has its place in space among the material things of the external world, and by which I am in communication with other similar individuals. Consciousness, without surrendering its immanence, becomes a part of reality, becomes this particular person, myself, who was born and will die. Moreover, as a result of this, consciousness spreads out its web, in the form of time, over reality. Change, motion, elapse of time, coming and ceasing to be, exist in time itself; just as my will acts on the external world through and beyond my body as a motive power, so the external world is in its turn active. Its phenomena are related throughout by a causal connection. In fact, physics shows that cosmic time and physical time cannot be disassociated from one another. The new solution of the problem of amalgamating space and time offered by the theory of relativity brings with it a deeper insight into the harmony of action in the world. (*ibid.*, 6)

Thus, for Weyl, the duality between mind and material reality leads to a unity between space and time.

Weyl believed that, in the nature of things, science was ultimately limited in its capacity to describe the world. Thus in the Preface to his luminous work *The Open World* (1932) he expresses the opinion that:

> Modern science, insofar as I am familiar with it through my own scientific work, mathematics and physics make the world appear more and more an open one. ... Science finds itself compelled, at once by the epistemological, the physical and the constructive-mathematical aspect of its own methods and results, to recognize this situation. It remains to be added that science can do no more than show us this open horizon; we must not by including the transcendental sphere attempt to establish anew a closed (though more comprehensive) world. (Weyl, *Open World*, v)

And in his *Address on the Unity of Knowledge* (1954) he says, in a particularly striking passage lightened by a touch of humour:

> The riddle posed by the double nature of the ego certainly lies beyond those limits [i.e., of science]. On the one hand, I am a real individual man, born by a mother and destined to die, carrying out real and psychical acts (far too many, I may think, if boarding a subway during rush hour). On the other hand , I am "vision" open to reason, a self-penetrating light, immanent sense-giving consciousness, or how ever you may call it, and as such unique. (Weyl, *Address*, 3)

In the same address, he reiterates that

> ...at the basis of all knowledge lies, firstly, intuition, mind's ordinary act of seeing what is given to it. (*ibid.*, 7)

In particular, Weyl held, like Brouwer, that intuition, or *insight*—rather than *proof*—furnishes the ultimate foundation of *mathematical* knowledge[3]. Thus in *The Continuum* (1918) he says, with an unusual hint of asperity,

> In the Preface to Dedekind (1888) we read that "In science, whatever is provable must not be believed without proof." This remark is certainly characteristic of the way most mathematicians think. Nevertheless, it is a preposterous principle. As if such an indirect concatenation of grounds, call it a proof though we may, can awaken any "belief" apart from assuring ourselves through immediate insight that each individual step is correct. In all cases, this process of confirmation—and not the proof—remains the ultimate source from which knowledge derives its authority; it is the "experience of truth". (Weyl, *Continuum*, 119)

Weyl recognized, however, that requiring all mathematical knowledge to possess intuitive immediacy is an unattainable ideal:

> The states of affairs with which mathematics deals are, apart from the very simplest ones, so complicated that it is practically impossible to bring them into full givenness in consciousness and in this way to grasp them completely. (*ibid.*, 17)

But Weyl did not think that this fact furnished justification for extending the bounds of mathematics to embrace notions which cannot be given intuitively *even in principle*

---

[3] It is to be noted in this connection that in *The Continuum* Weyl also makes the assertion that
> ...the idea of iteration, i.e., of the sequence of natural numbers, is an ultimate foundation of mathematical thought .... (p. 48)

(e.g., the actual infinite). He believed, rather, that this extension of mathematics into the transcendent—the world beyond immediate consciousness—is made necessary by the fact that mathematics plays an indispensable role in the physical sciences, in which intuitive evidence is *necessarily* transcended:

> ... if mathematics is taken by itself, one should restrict oneself with Brouwer to the intuitively cognizable truths ... nothing compels us to go farther. But in the natural sciences we are in contact with a sphere which is impervious to intuitive evidence; here cognition necessarily becomes symbolical construction. Hence we need no longer demand that when mathematics is taken into the process of theoretical construction in physics it should be possible to set apart the mathematical element as a special domain in which all judgements are intuitively certain; from this higher standpoint which makes the whole of science appear as one unit, I consider Hilbert to be right. (Weyl, *Open World*, 82)

In *Consistency in Mathematics* (1929), Weyl characterized the mathematical method as

> the *a priori* construction of the possible in opposition to the *a posteriori* description of what is actually given. (Weyl, *Consistency*, 249)

The problem of identifying the limits on constructing "the possible" in this sense occupied Weyl a great deal. He was particularly concerned with the concept of the mathematical *infinite*, which he believed to elude "construction" in the idealized sense of set theory:

> No one can describe an infinite set other than by indicating properties characteristic of the elements of the set. ... The notion that a set is a "gathering" brought together by infinitely many individual arbitrary acts of selection, assembled and then surveyed as a whole by consciousness, is nonsensical; "inexhaustibility" is essential to the infinite. (Weyl, *Continuum*, 23)

But the necessity of expanding mathematics into the external world forces it to embody a conception of the actual infinite, as Weyl attests towards the end of *The Open World*:

> The infinite is accessible to the mind intuitively in the form of a field of possibilities open to infinity, analogous to the sequence of numbers which can be continued indefinitely, but the completed, the actual infinite as a closed realm of actual existence is forever beyond its reach. Yet the demand for totality and the metaphysical belief in reality inevitably compel the mind to represent the infinite by closed being as symbolical construction. (Weyl, *Open World*, 83)

Another mathematical "possible" to which Weyl gave a great deal of thought is the idea of the *continuum*. During the period 1918–1921 he wrestled unceasingly with the problem of providing it with an exact mathematical formulation free of the taint of the actual infinite. As he saw it, there is an unbridgeable gap between intuitively given continua (e.g. those of time and motion) on the one hand, and the "discrete" exact concepts of mathematics (e.g. that of real number) on the other:

> ... the conceptual world of mathematics is so foreign to what the intuitive continuum presents to us that the demand for coincidence between the two must be dismissed as absurd (Weyl, *Continuum*, 108);

... the continuity given to us immediately by intuition (in the flow of time and of motion) has yet to be grasped mathematically as a totality of discrete "stages" in accordance with that part of its content which can be conceptualized in an exact way (*ibid.*, 24);

When our experience has turned into a real process in a real world and our phenomenal time has spread itself out over this world and assumed a cosmic dimension, we are not satisfied with replacing the continuum by the exact concept of the real number, in spite of the essential and undeniable inexactness arising from what is given. (*ibid.*, 93).

By 1921 he had joined Brouwer in regarding the continuum as a "medium of free becoming"; in one of his last papers he observes:

... constructive transition to the continuum of real numbers is a serious affair... and I am bold enough to say that not even to this day are the logical issues involved in that constructive concept completely clarified and settled. (Weyl, *Axiomatic*, 17)

It is greatly to be regretted that Weyl did not live to see the startling developments stemming from nonstandard analysis and, especially, smooth infinitesimal analysis, which provides a mathematical framework within which the concept of a true continuum—that is, not "synthesized" from discrete elements—can be formulated and developed (*q.v.* Appendix 3).

It seems appropriate to conclude with a quotation from *The Open World* which expresses with clarity and eloquence the essence of Weyl's philosophy:

The beginning of all philosophical thought is the realization that the perceptual world is but an image, a vision, a phenomenon of our consciousness; our consciousness does not directly grasp a transcendental real world which is as it appears. The tension between subject and object is no doubt reflected in our conscious acts, for example, in sense perceptions. Nevertheless, from the purely epistemological point of view, no objection can be made to a phenomenalism which would like to limit science to the description of what is "immediately given to consciousness". The postulation of the real ego, of the thou and of the world, is a metaphysical matter, not judgment, but an act of acknowledgment and belief. But this belief is after all the soul of all knowledge. It was an error of idealism to assume that the phenomena of consciousness guarantee the reality of the ego in an essentially different and somehow more certain manner than the reality of the external world; in the transition from consciousness to reality the ego, the thou and the world rise into existence indissolubly connected and, as it were, at one stroke. (Weyl, *Open World*, 26–27)

BIBLIOGRAPHY

Alexandrov, A.D., A.N.Kolmogorov, and M.A.Lavrentiev (eds.), *Mathematics, its Content, Methods and Meaning*, 3 vols., tr. S.H. Gould and T. Bartha. Cambridge, Mass.: MIT Press, 1969.

Aristotle, *Physics*, 2 vols., tr. P. H. Wickstead and F. M. Cornford. Loeb Classical Library, Cambridge, Mass.: Harvard University Press and London: Heinemann, 1980.

Baron, M.E., *The Origins of the Infinitesimal Calculus*. New York: Dover, 1987.

Bell, E.T., *The Development of Mathematics*, 2nd ed. New York: McGraw-Hill, 1945.

~~~ *Men of Mathematics*, 2 vols. Harmondsworth, Middlesex: Penguin Books, 1965.

Bell, J.L., *A Primer of Infinitesimal Analysis*. Cambridge: Cambridge University Press, 1998.

Benacerraf, P., and H. Putnam, *Philosophy of Mathematics: Selected Readings*, 2nd ed. Cambridge: Cambridge University Press, 1977.

Bernardete, J., *Infinity: An Essay in Metaphysics*. Oxford: Clarendon Press, 1964.

Binmore, K.G., *Mathematical Analysis: A Straightforward Approach*. Cambridge: Cambridge University Press, 1977.

Birkhoff, G., and S. Mac Lane, *A Survey of Modern Algebra*. New York: Macmillan, 1959.

Bix, R., *Conics and Cubics: A Concrete Introduction to Algebraic Curves*. New York Springer-Verlag, 1998.

Black, M., *The Nature of Mathematics*. London: Routledge and Kegan Paul,.1958.

Bochner, S., *The Role of Mathematics in the Rise of Science*. Princeton, N.J.: Princeton University Press, 1966.

Bonola, R., *Non-Euclidean Geometry*. New York: Dover, 1955.

Borel, E., *Space and Time*. New York: Dover, 1960.

Bourbaki, N., *Elements of the History of Mathematics*, tr. J. Meldrum. Berlin: Springer- Verlag, 1994.

Boyer, C., *A History of Mathematics*, 2nd ed. New York: Wiley, 1991.

~~~ *The History of the Calculus and its Conceptual Development*. New York: Dover, 1969.

Cajori, F., *A History of Mathematical Notations*. New York: Dover, 1993.

Cantor, G., *Contributions to the Founding of the Theory of Transfinite Numbers*. New York: Dover, 1961.

Cardan, J., *The Book of My Life*, tr. J. Stover. New York: Dover, 1962.

Coe, M., *The Maya*. Harmondsworth, Middlesex: Penguin Books, 1973.

Coolidge, J.L., *A History of the Conic Sections*. New York: Dover, 1968.

Courant, R., *Differential and Integral Calculus*, 2 vols., tr. E.J. McShane. London and Glasgow: Blackie, 1942.

~~~ and H. Robbins, *What is Mathematics?* New York: Oxford University Press, 1967.

Dantzig, T., *Number, The Language of Science*. New York: Doubleday, 1954.

Dedekind, R., *Essays on the Theory of Numbers*, tr. W.W. Beman. Chicago: Open Court, 1901.

DeLong, H., *A Profile of Mathematical Logic*. Reading, Mass.: Addison-Wesley, 1970.

Demopoulos, W. (ed.), *Frege's Philosophy of Mathematics*. Cambridge, Mass.: Harvard University Press, 1995.

Dudley, U., *Elementary Number Theory*, 2nd ed. San Francisco: W.H. Freeman, 1978.

Dummett, M., *Elements of Intuitionism*. Oxford: Clarendon Press, 1977.

Einstein, A., *Relativity: the Special and General Theory*. London: Methuen, 1970.

Eves, H., *A History of Mathematics*, 4th ed. New York: Holt, Rinehart and Winston, 1976.

Farrington, B., *Greek Science*, 2 vols. Harmondsworth, Middlesex: Penguin Books, 1949.

Fraenkel, A., Y. Bar-Hillel and A. Levy, *Foundations of Set Theory*. Amsterdam: North-Holland, 1973.

Frege, G., *The Foundations of Arithmetic*, tr. J.L. Austin. Evanston, Ill.: Northwestern University Press, 1980.

~~~ *The Basic Laws of Arithmetic*, tr. M. Furth. Berkeley, Ca.: University of California Press, 1967.

Gamow, G., *One, Two, Three, ... Infinity: Facts and Speculations of Science*. New York: New American Library, 1957.

Gray, J., *Ideas of Space: Euclidean, Non-Euclidean, and Relativistic*. Oxford: Clarendon Press, 1973.

Greenberg, M., *Euclidean and Non-Euclidean Geometry*. San Francisco: W.H. Freeman, 1974.

Grunbaum, A., *Zeno's Paradoxes and Modern Science*. London: Allen and Unwin, 1967.

Hadamard, J., *The Mathematician's Mind*. Princeton, N.J.: Princeton University Press, 1996.

Hallett, M., *Cantorian Set Theory and Limitation of Size*. Oxford: Clarendon Press, 1984.

Hardy, G.H., *A Mathematician's Apology*. Cambridge: Cambridge University Press, 1960.

~~~   *A Course in Pure Mathematics*. Cambridge: Cambridge University Press, 1960.

~~~ and E.M. Wright, *The Theory of Numbers*. Oxford: Clarendon Press, 1938.

Heath, T., *A History of Greek Mathematics*, 2 vols. New York: Dover, 1981.

Heijenoort, J. van, *From Frege to Gödel: A Sourcebook in Mathematical Logic 1879 – 1931*. Cambridge, Mass.: Harvard University Press, 1981.

Heyting, A., *Intuitionism: An Introduction*. Amsterdam: North-Holland, 1971.

Hilbert, D., and S. Cohn-Vossen, *Geometry and the Imagination*. New York: Chelsea, 1952.

Hocking, J.G. and G.S. Young, *Topology*. Reading, Mass.: Addison-Wesley, 1961.

Ivins, W.M., *Art and Geometry*. New York: Dover, 1964.

Jacobson, N., *Basic Algebra I*. New York: W.H. Freeman, 1985.

Kant, I., *Critique of Pure Reason*. London: Macmillan, 1964.

Kelley, J.L., *General Topology*. Princeton, N.J.: Van Nostrand, 1960.

Kirk, G.S., J. E. Raven, and M. Schofield, *The Presocratic Philosophers*, 2nd ed. Cambridge: Cambridge University Press, 1983.

Klein, J., *Greek Mathematical Thought and the Origins of Algebra*, tr. E. Braun. New York: Dover, 1992.

Kline, M., *Mathematics in Western Culture*. Harmondsworth, Middlesex: Penguin Books, 1972.

~~~   *Mathematical Thought from Ancient to Modern Times*. New York: Oxford University Press, 1972.

~~~   *Mathematics: The Loss of Certainty*. New York: Oxford University Press, 1980.

Kneale, W., and M. Kneale, *The Development of Logic*. Oxford: Clarendon Press, 1991.

Kneebone, G.T., *Mathematical Logic and the Foundations of Mathematics*. London: Van Nostrand, 1965.

Körner, S., *Philosophy of Mathematics*. London: Hutchinson, 1960.

Lakatos, I., *Proofs and Refutations*. Cambridge: Cambridge University Press, 1976.

Lawvere, F.W., and S. Schanuel, *Conceptual Mathematics: A First Introduction to Categories*. Cambridge: Cambridge University Press, 1997.

Lionnais, F. le, (ed.), *Great Currents of Mathematical Thought*, 2 vols., tr. R. Hall. New York: Dover, 1971.

Lipschutz, M., *Differential Geometry*. New York: McGraw-Hill and Schaum's Outline Series, 1969.

Machover, M., *Set Theory, Logic and Their Limitations*. Cambridge: Cambridge University Press, 1976.

Mac Lane, S., *Mathematics, Form and Function*. New York: Springer-Verlag, 1986.

~~~ and G. Birkhoff, *Algebra*. London: Macmillan, 1967.

Mancosu, P., *From Brouwer to Hilbert: The Debate on the Foundations of Mathematics in the 1920s*. New York: Oxford University Press, 1998.

Marsden, J.E., *Basic Complex Analysis*. San Francisco, W.H. Freeman, 1973.

~~~   *Elementary Classical Analysis*. San Francisco: W.H. Freeman, 1974.

~~~ and Tromba, A.J., *Vector Calculus*, 2nd ed. San Francisco: W.H. Freeman, 1981.

Meschkowski, H., *Evolution of Mathematical Thought*, tr. J.H. Gayl. San Francisco: Holden-Day, 1965.

Moore, A.W., *Infinity*. London: Routledge, 1990.

Moore, G.H., *Zermelo's Axiom of Choice: Its Origins, Development, and Influence.*. New York: Springer-Verlag, 1982.

Needham, J., and Wang Ling, *Science and Civilization in China, vol. 3: Mathematics and the Sciences of the Heavens and Earth*. Cambridge: Cambridge University Press, 1959.

Newman, J.R.(ed.), *The World of Mathematics*, 4 vols. New York: Simon and Schuster, 1956.

Newton, I., *Principia*, tr. A. Motte, revsd. F. Cajori. Berkeley, Ca.: University of California Press, 1962.

Phillips, E.G., *Functions of a Complex Variable*. Edinburgh and London: Oliver and Boyd, 1961.

Plato, *The Collected Dialogues*, E. Hamilton and H. Cairns, eds. Princeton, N.J.: Princeton University Press, 1961.

Reichenbach, H., *Philosophy of Space and Time*. New York: Dover, 1959.

Rescher, N., *The Philosophy of Leibniz*. Englewood Cliffs, N.J.: Prentice-Hall, 1967.

Robinson, A., *Non-Standard Analysis*. Princeton, N.J.: Princeton University Press, 1996.

Rotman, B., *Signifying Nothing: The Semiotics of Zero*. Palo Alto, Ca.: Stanford University Press, 1987.

~~~ *Ad Infinitum: The Ghost in Turing's Machine*. Palo Alto, Ca.: Stanford University Press, 1993.

Russell, B., *The Principles of Mathematics*. London: Allen and Unwin, 1964.

~~~ *Introduction to Mathematical Philosophy*. London: Routledge, 1995.

~~~ *Mysticism and Logic*. Harmondsworth, Middlesex: Penguin Books, 1953.

~~~ and A.N. Whitehead, *Principia Mathematica to *56*. Cambridge, Cambridge University Press, 1962.

Sambursky, S. *The Physical World of the Greeks*. London: Routledge and Kegan Paul, 1963.

~~~ *Physics of the Stoics*. London: Hutchinson, 1971.

Smith, D.E., *A Sourcebook in Mathematics*. New York: Dover, 1959.

Struik, D., *A Concise History of Mathematics*. New York: Dover, 1948.

Sondheimer, E., and A. Rogerson, *Numbers and Infinity: An Historical Account of Mathematical Concepts* Cambridge: Cambridge University Press, 1981.

Szabo, A., *The Beginnings of Greek Mathematics*, tr. A.M. Ungar. Hingham, Mass.: D. Reidel, 1978.

Titchmarsh, E.C., *Theory of Functions*, 2nd ed. London: Oxford University Press, 1958

van der Waerden, B.L., *Modern Algebra*, 2 vols., tr. F. Blum and T. Benac. New York: Ungar, 1953.

Wagon, S., *The Banach-Tarski Paradox*. Cambridge: Cambridge University Press, 1993.

Weyl, H., *Philosophy of Mathematics and Natural Science*. Princeton: Princeton University Press, 1949.

~~~ *Symmetry*. Princeton: Princeton University Press, 1989.

~~~ *The Continuum: A Critical Examination of the Foundation of Analysis*, tr. S. Pollard and T. Bole. Kirksville, Mo.: Thomas Jefferson University Press, 1987.

~~~ *Space-Time-Matter*, tr. H.L. Brose. New York: Dover, 1950.

~~~ *The Open World: Three Lectures on the Metaphysical Implications of Science*. New Haven, Ct.: Yale University Press, 1932.

~~~ "Address on the Unity of Knowledge." Columbia University Bicentennial Celebration, 1954.

~~~ "Axiomatic versus Constructive Procedures in Mathematics", ed. T . Tonietti. *Mathematical Intelligencer* 7, no. 4, 10–17, 38 (1985).

~~~ "Consistency in Mathematics." Rice Institute Pamphlet 16, 245–265 (1929).

White, M.J., *The Continuous and the Discrete*. Oxford: Clarendon Press, 1992.

Whitehead, A.N., *Science and the Modern World*. Cambridge: Cambridge University Press, 1953.

Wilder, R., *Introduction to the Foundations of Mathematics*. New York: Wiley, 1952.

Zippin, L., *Uses of Infinity*. New York: Random House, 1962.

INDEX OF NAMES

INDEX OF TERMS

The Western Ontario Series
in Philosophy of Science

The Western Ontario Series
in Philosophy of Science

The Western Ontario Series
in Philosophy of Science

The Western Ontario Series
in Philosophy of Science

61. M.C. Galavotti and A. Pagnini (eds.): *Experience, Reality, and Scientific Explanation.* 1999
ISBN 0-7923-5497-4
62. D. Fisette (ed.): *Consciousness and Intentionality: Models and Modalities of Attribution.* 1999
ISBN 0-7923-5907-0
63. J.L. Bell: *The Art of the Intelligible.* An Elementary Survey of Mathematics in its Conceptual
Development. 1999   ISBN 0-7923-5972-0